Organic Fertilizers - Their Role in Sustainable Agriculture

Edited by Manuel Oliveira

Published in London, United Kingdom

Organic Fertilizers - Their Role in Sustainable Agriculture
http://dx.doi.org/10.5772/10.5772/intechopen.1005724
Edited by Manuel Oliveira

Contributors
Agief Julio Pratama, Alexander Kurovsky, Andrey Babenko, Betsabé Rodríguez Pérez, Clement Oppong Peprah, Daniela Piaz Barbosa Leal, Edi Wiraguna, Ekaterina Kornievskaya, Eric Owusu Danquah, Fabián Pérez-Labrada, Farzaneh Abdolmaleki, Felix Frimpong, Henry Kasman Hadi Saputra, Hermilo Lucio Castillo, Imran Khan, Irene Iliana Ramírez-Bustos, Israel Ogwuche Ogra, Jeannette Aduhene-Chinbuah, José Alfonso López García, José Antonio Gama-Salgado, Julieta Grajales Conesa, Julio Javier Gastelum Reyes, Karla Ivette Chy-young Peñuñuri, Kennedy Agyeman, Kwaku Onwona-Hwesofour Asante, Leopoldo Caridad Cruz López, Lorena Alemán-Figueroa, Manuel T. Oliveira, Maria Guadalupe De Gyves Córdova, Maribel López-Atanacio, Mari Carmen López-Pérez, Marycarmen Utrilla Vázquez, Milad Daneshniya, Miriam Mendivil Morales, Mohammad Hossein Maleki, Mohammed Sani Jibril, Muhammad Iqbal Nurulhaq, Nicolai David Jablonowski, Otávio dos Anjos Leal, Ralph Bam, Ratih Kemala Dewi, Restu Puji Mumpuni, Ririh Sekar Mardisiwi, Rodolfo Torres de los Santos, Rodrigo Lucas-García, Stephan Blossfeld, Tri Budiarto, Víctor Jesús Albores Flores, Víctor Rosas-Guerrero, Widya Hasian Situmeang, Yana Gummer, Yana Gummer

First published in London, United Kingdom, 2025 by IntechOpen
IntechOpen is the global imprint of INTECHOPEN LIMITED, registered in England and Wales, registration number: 11086078, 167-169 Great Portland Street, London, W1W 5PF, United Kingdom

For EU product safety concerns: IN TECH d.o.o., Prolaz Marije Krucifikse Kozulić 3, 51000 Rijeka, Croatia, info@intechopen.com or visit our website at intechopen.com.

British Library Cataloguing-in-Publication Data
A catalogue record for this book is available from the British Library

Organic Fertilizers - Their Role in Sustainable Agriculture
Edited by Manuel Oliveira
p. cm.

This title is part of the Agricultural Sciences Book Series, Volume 21
Topic: Agronomy and Horticulture
Series Editor: W. James Grichar
Topic Editor: Ibrahim Kahramanoglu
Associate Topic Editors: Murat Helvaci and Olga Panfilova

Print ISBN 978-1-83634-416-2
Online ISBN 978-1-83634-415-5
eBook (PDF) ISBN 978-1-83634-417-9
ISSN 3029-052X

If disposing of this product, please recycle the paper responsibly.

IntechOpen Book Series

Agricultural Sciences

Volume 21

Aims and Scope of the Series

The importance of agriculture cannot be overstated. It helps sustain life, as it gives us the food we need to survive and provides opportunities for economic well-being. Agriculture helps people prosper around the world and combines the creativity, imagination, and skill involved in planting crops and raising animals with modern production methods and new technologies. This series includes two main topics: Agronomy and Horticulture, and Animal Farming. This series will help readers better understand the intricacies of production agriculture and provide the new knowledge that is required to be successful. The success of a farmer in modern agriculture requires knowledge of events happening locally as well as globally that impact input decisions and ultimately determine net profit.

Meet the Series Editor

W. James Grichar has been employed with Texas A&M AgriLife Research for over 45 years with an emphasis on research in agronomy, plant pathology, and weed science. He obtained his BS from Texas A&M in 1972 and his Masters of Plant Protection in 1975. He has published 195 journal articles, over 330 research reports and briefs, 11 book chapters, and over 300 abstracts of profession meetings. He also directs research in many crops including corn, grain sorghum, peanuts, and sesame. He has held various positions in different professional societies including the American Peanut Research and Education Society, Southern Weed Science Society, and Texas Plant Protection Conference in addition to being Associate Editor for Peanut Science and Weed Technology. Significant accomplishments have included spearheading efforts to determine the optimum planting time for soybean production along the upper Texas Gulf Coast. These efforts have shown growers that soybean yields can be improved by 10 to 20% by following a late March to early April plant date. He also has been instrumental in developing a herbicide program for peanut production in the south Texas growing region. Through the development and use of herbicides that are effective against major weed problems in the south Texas region, peanut yields have increased by 25 to 30%.

Meet the Volume Editor

Manuel T. Oliveira is a retired professor from the Department of Agronomy at the University of Trás-os-Montes and Alto Douro (UTAD) in Portugal, specializing in soil science, agricultural systems, and viticulture. He was previously the head of the Department of Agronomy and a member of the scientific council of UTAD. He was an invited researcher at Cornell University (USA), Catholic University of Santiago (Chile), and Universidad de La Rioja (Spain). He is a member of the editorial council of Plant and Soils and a regular referee for Sustainable Agriculture Research and Agriculture, Ecosystems & Environment. He is a member of the Horticultural Society of Portugal, the Portuguese Society of Agricultural Science, and the Portuguese Association for National Defense. He is a scientific committee member of the 2nd Global Summit on Climate Change & Environmental Sustainability.

Contents

Preface

This book discusses the role of organic fertilizers in plant nutrition and sustainable agriculture. It highlights the importance of plant nutrition in coping with environmental and climate-related stresses, ensuring food security, and maintaining soil health. Key points include:

1. Plant Nutrition & Fertilization: Essential nutrients are required for plant growth, and fertilizers help meet these needs. However, excessive chemical fertilizer use leads to environmental pollution and sustainability concerns.

2. Challenges of Chemical Fertilizers: Rising costs due to energy market fluctuations, geopolitical instability, and their negative impact on soil microbiomes and ecosystems.

3. Organic Fertilizers: Sourced from compost, manure, biochar, and microbial inoculations, they enhance soil health and reduce dependence on synthetic chemicals but have limitations in providing sufficient nutrients.

4. Benefits & Risks: Organic fertilizers improve soil structure and fertility but may introduce contaminants like heavy metals, pathogens, and antibiotic residues. Composting and biochar have potential but require further research.

5. Future Directions: A balanced approach combining organic and chemical fertilizers, along with microbial biofertilizers, could enhance sustainability while maintaining productivity.

Manuel T. Oliveira
University of Trás-os-Montes and Alto Douro,
Vila Real, Portugal

Section 1

Introduction

Chapter 1

Introductory Chapter: Organic Fertilizers and Plant Nutrition – Devising Ways to Healthier Agricultural Systems

Manuel T. Oliveira

1. Introduction

Scientific developments in agriculture and technologies of chemical fertilizers and pesticides fed the "green revolution" of mid-twentieth century, but few decades later pollution and toxics from those chemicals became evident. Now it is climate change, partially caused by agricultural technologies, that have also moved to the center of our preoccupations. These environmental problems, but also economic and social inequities, incentive the search for more sustainable agricultural technologies that can be brought about by deeper scientific insight. Replacing chemical fertilizers by less harmful products, lets call them organic fertilizers, and still maintain crop productions capable of feeding the world population is an objective for farmers, policy makers, and, in fact, for everybody else. In today world science and technology move forward at very rapid pace, pervading every aspect of social and individual lives; keeping in touch with them is a necessity for each one of us in our field of work.

2. Organic fertilizers: Why to use them?

Knowledge of plant nutrition and its capacity to adapt and cope with biotic and abiotic stresses such as environmental changes are essential to achieve high levels of crop production and to adequate management of diverse ecosystems. Plant nutrition is one of the plant sciences that deals with plant productivity that is a major objective of agriculture systems with a significant impact on global food security and it has a direct and indirect relationships with human health. Plant nutrition includes the study of plant growth and development, the supply of nutrients on the right amount, time, and source given the prevailing environmental conditions. The paradigm of plant nutrition always accompanied the evolution of scientific knowledge but now there is an urgency regarding this science as the rapid increase of stressful conditions created by climate change might endanger the world food security. New insights on plant nutrition and on how to supply nutrients may help reduce the negative impacts of abiotic stresses and associated biotic ones and prevent our production systems to suffer a large reduction of productivity.

Plant nutrition needs nutrients which are indispensable for growth and development and the plant nutritional state affects their quality as food and feed. Proper nutrient management also helps the fight against plant diseases as the host susceptibility and the pathogen's virulence are handled according to the necessities. Plants require 16 essential mineral nutrients [1] for their growth and development as they are constituents of nucleic acids, phospholipids, specific amino acids, various co-enzymes, among many other molecules, indispensable for cellular metabolism [2, 3]. Of course, plants use many other mineral nutrients that are not deemed essential but they can play an important role on the plant metabolism. The traditional definition of essentiality might not represent entirely the most actual concepts of plant nutrition, the agronomic practices and the regulatory framework applied to agriculture.

A proposed new definition of plant nutrient might read as "*A mineral plant nutrient is an element which is essential or beneficial and development or for the quality attributes of the harvested product of a given plant species grown in its natural or cultivated environment. A plant nutrient may be considered essential if the life cycle of a diversity of plant species cannot be completed in the absence of the element. A plant nutrient may be considered beneficial if it does not meet the criteria of essentiality, but can be shown to benefit plant growth and development or the quality attributes of a plant or its harvested product*" [4].

Plant nutrients are mostly obtained from the soil or the soil solution but one or more of the needed mineral elements might not have the required level of phytoavailability (or it can be in toxic concentration), thus resulting in limited crop production. Where the availability is insufficient, agricultural crops are supplied with fertilizers containing not only essential but also beneficial nutrients for better edible organs and then beneficial to human consumption. Nowadays, fertilizers are indispensable to reach the amount and quality of produce to feed the ever-increasing world population. However, the nutritional efficiency of the fertilizers (the proportion of nutrients from fertilizers that actually contribute to plant growth and development) might be low due to losses by leaching, insolubility, decomposition, etc. Hence, farmers resort to the excessive amount of fertilizers and a large part of them dissipate into the atmosphere, surface and sub-surface water bodies where they cause serious environmental pollution.

Facing a more stressful environment, pollution problems, loss of biodiversity, world agriculture has to transition to sustainable production systems using lower amounts of chemical inputs and preserve the diversity of microbiomes as plants in association with microorganisms are considered "holobionts" an organism consisting of a macrobe and its microbial associates [5, 6]. Soil microbes regulate the functions and cycles of soil ecosystems, they are indispensable in several process of element cycling, and pollutant degradation; their diversity and biomass reflect the conditions of the environment such as soil nutrients, pH, and other external conditions [7, 8]. Better understanding of association plant-microorganism might be a useful tool of actual agriculture in dealing with our changing environment.

Fertilization is one of the most important production factors responsible for increased yields and improved quality of crops. The lack of adequate fertilization would lead to a decrease in agricultural productivity and consequent higher food costs. However, the market instability of fossil fuels and phosphorus supply has created an inflationary tendency on the costs of fertilizers during the last decades and may continue in the future, raising doubts about the possibility of crop intensification to meet a growing food demand [9]. The Russian-Ukrainian crisis is another factor of instability on the fertilizer market and their prices have increased by 30% since the beginning of the war in Ukraine [10].

The most used fertilizers on farms include nitrogen, potassium, and phosphate fertilizers and in 2022 the world demand of was over 200,000 thousand metric tons [11, 12]. Production of chemical fertilizers is now a major worldwide industry that sustained the 'Green Revolution' still considered an important progress that permitted to raise far more crops than was previously possible, but the dark side of chemical pollution of soils and water from leaching and runoff from agriculture fields has raised serious concerns all over the world. Ammonium fertilizers are by far the main and most common; they sustain higher yields that make the backbone of farmer's profitability. However, it is well documented that nitrogen excess pollutes the atmosphere with nitrogen compounds, cause acid rain, and nitrate (NO_3) leakages to aquatic ecosystems [13, 14].

These are some of the problems facing the agricultural systems and the reasons farmers must become less dependent on synthetic resources that might affect negatively the soil microbiome but still maintain adequate levels of productivity. Many countries promote actively agriculture strategies relying less on chemical products and one example is "The European Green Deal" policies (Available from: https://commission.europa.eu/strategy-and-policy/priorities-2019-2024/european-green-deal_en) calling for more "*sustainable and environmentally conscious agricultural practices*", reducing the use of chemical fertilizer and pesticides. We are well aware that life is a symbiotic interplay between mutualism and parasitism controlled by environmental conditions and if life is to thrive people must be fed, food must have good quality and environment has to be protected.

2.1 Organic fertilizers: What are they?

Organic fertilizers are part of the response to global need of reduced use of chemical compounds. An organic fertilizer is mainly sourced from peat or from fermentation and decomposition of agricultural wastes, both plant and animal husbandry. Other materials like biochar, a wide range of living organisms applied as biofertilizers, powdered rocks such as lime, rock phosphate, and greensand, thou not of biological organs, can be considered in the same category. These materials break down naturally and provide nutrient and minerals to the soil needed for plant growth and, at least conceptually, they might sustain and enhance the health of soil, plants, animals and humans. Organic fertilizers can also be very attractive to the farmer because some of them can be produced locally or even on the farm itself, reducing the costs, promoting a circular economy.

The World Bank estimates that the overall waste production will increase to 3.40 billion metric tons by 2050. Only about 14% of waste is recycled and 6% is composted, therefore, waste recycling is a pressing need and a suitable method is to turn waste materials into fertilizers [15]. Organic wastes are from a very diverse origin: animal organic waste (manure), compost (plant, food waste), urban waste (sewage sludge), and others, each have different characteristics and they are processed (inoculation with microorganisms, acid leaching, anaerobically digested, pyrolysis, etc.) to optimize their nutrients content and promote their agricultural value.

Animal manures as fertilizer has been carried out across the globe for centuries. They contain a variety of organic acids, peptides, and nutrients NPK and micronutrients to produce high yielding and good quality crops. However, use of manures is not problem free. Long-term and heavy application of manure application might lead to soil salinization under non-irrigated conditions and leaching of soluble salts to the groundwater under irrigated conditions [16]. Manure may contain infectious agents and toxic chemicals, like antibiotics that can be introduced to the food chain. Organic fertilizers contaminated by human and animal pathogens have been the cause of food-borne epidemics. Antibiotics and other chemicals used by humans and in animal feeds will also end up in the organic fertilizers and they will enhance the occurrence of resistant and multi-resistant strains of microorganisms in both human and animal populations; there are also reports of ecotoxicological environmental effects and disruption of the ecological balance caused by them. Exposure of microorganisms to sublethal concentration of antibiotics in the organic products induces antibiotic resistance [17]. Heavy metal enrichment of soil and imbalanced nutrient availability might also occur [18]. Manures can cause nutrient losses to water and the atmosphere due to priming effects and their promotion of nitrification and denitrification [19].

It is common to use the term compost and fertilizer interchangeably but we should not make that mistake. Compost lacks some nutrients or they are present in too low concentration to feed crop adequately. But compost ameliorates the physical condition of the soil and in this sense, it is a soil amendment rather than a fertilizer. Because compost is decomposed organic matter and improves the quality of soil, it is sometimes included in the category of organic fertilizer. Adding compost to soil results in sustained increases in organic matter pools that brings clear benefits to soil in form of lower bulk density, better water retention in coarser soils, higher microbial activity and slow release of nutrients preventing rapid losses [20, 21].

Composting can reduce the volume of organic wastes in an environmentally friendly way and it might be economically viable as it cuts costs related to hauling of wastes and farmers to reduce the use of chemical fertilizers. As we try to mitigate global warming, composting holds the promise of one of the options available, benefiting societies environmentally, economically, and in overall health [22]. Reusing organic wastes is a form of recovering resources, a tenant of circular economy [23].

Nevertheless, there are concerns about the use of compost as soil amendment. Plastic pollution, a global threat to environmental and human health, is present in soils, including agricultural soils, receiving urban compost amendments [24]. Antibiotic-resistant bacteria, selected by antibiotic use in livestock, are spreading to crops from wastes via composting and represent a public health threat as they find their way into the food chain and transmitted to humans [25]. Compost from sewage treatment plants contain organic materials and nutrients useful for crops but they can also be harmful if contain organic pollutants, heavy metals, or pathogens [26]. Sewage compost, in general, contain more heavy metals (Zn, Cu, Ni, Pb and Cd) than the background presence in the soil; its continuous application as amendment will increase the heavy metal concentration in soil and, in lesser extent, in crops [27, 28].

Biochar is a solid carbon material produced by pyrolysis of biomass under high temperature and limited oxygen. Its use as fertilizer has been reported to improve soil structure, carbon storage, water and nutrient retention, increase productivity and crop profitability. Biochar can improve soil quality by reducing bulk density and increasing the soil retention capacity for water. It shows a potential for mitigating CO_2 emissions from the cropping system. Its application to grain crop might improve yields only slightly but its overall impact is possibly positive [29]. There is already a large body of research on biochar but some important gaps in knowledge still remain, such as no standard characterization of chemical and physical properties of biochar and no long term effects of its effect on soil is available [30]. Despite all its potential benefits, they cannot always be optimized and some disadvantages have been found [31]. Examples reported say that they reduce pH in acidic soils [32] but also cause nutrient precipitation and reduce plant utilization of nutrients in alkaline soils [33];

it increases soil water retention in coarse soils [34] but reduces the water holding capacity in clay soils [35].

Microbial inoculations are used for biofortification, biofertilizers and biopesticides, which, apparently, have no damaging effects on the ecosystem [36]. Biofertilizer based on microorganisms comprises living microorganisms in microbial aggregates such as periphyton [8]. The microorganisms may be heterotrophic, photoautotrophic, or a complex mixture, including algae, bacteria, protozoa, metazoan and epiphytes [37]. In general, the term biofertilizer describes soil live microorganisms that enhances plant growth and development by a number of processes: more efficient nutrient uptake, lower losses of nutrients, excretion of phytohormones promoting plant growth, controlling pathogens, increasing plant resistance to biotic and abiotic stress, and pollutant detoxification [38, 39]. Inoculating the soil with bacteria capable of solubilizing phosphates is a promising strategy to improve plant phosphate uptake in P-deficient soils [40]. Plant growth-promoting bacteria have been found to increase available P and other nutrients, regulate plant stress responses, and relieving toxicity stresses by blocking and sequestering ions [41]. The genera *Azospirillum*, *Bacillus*, *Pseudomonas*, *Herbaspirillum* and *Burkholderia* have been extensive studied and they show capacity to increase plant nutrition, tolerance to pathogens and to stressful climate conditions [42].

It is not reasonable to expect that organic fertilizers alone can support plant high productivity since they do not provide enough NPK, thus, a combined addition of chemical and organic fertilizers might be a better solution, particularly for soils with low in N, P and organic C. This combination and the use of plant-growth-promoting microorganisms can lead to a significant decrease in chemical fertilizers demand [43, 44]. Today, we have a number of options to provide nutrients to the crops that can cause less environment harm, mitigate climate change and still maintain crop productivity for an ever-increasing population. To understand the inner mechanisms of such option is crucial to better manage our agriculture systems.

Author details

Manuel T. Oliveira
UTAD, Vila Real, Portugal

*Address all correspondence to: mto@utad.pt

References

[1] Tripathi R et al. Plant mineral nutrition and disease resistance: A significant linkage for sustainable crop protection. Frontiers in Plant Science. 2022;**13**:883970

[2] Soares JC et al. Preserving the nutritional quality of crop plants under a changing climate: Importance and strategies. Plant and Soil. 2019;**443**(1-2):1-26

[3] Elbasiouny H et al. Plant nutrition under climate change and soil carbon sequestration. Sustainability. 2022;**14**(2):914-934

[4] Brown PH, Zhao F-J, Dobermann A. What is a plant nutrient? Changing definitions to advance science and innovation in plant nutrition. Plant and Soil. 2021;**476**(1-2):11-23

[5] Vandenkoornhuyse P et al. The importance of the microbiome of the plant holobiont. New Phytologist. 2015;**206**(4):1196-1206

[6] Hacquard S. Disentangling the factors shaping microbiota composition across the plant holobiont. The New Phytologist. 2016;**209**(2):454-457

[7] Su J et al. Zokor disturbances indicated positive soil microbial responses with carbon cycle and mineral encrustation in alpine grassland. Ecological Engineering. 2020;**144**:105702

[8] Zhang Z-W et al. Nitrogen and phosphorus additions interactively affected composition and carbon budget of soil nematode community in a temperate steppe. Plant and Soil. 2022;**473**(1):109-121

[9] Brunelle T et al. Evaluating the impact of rising fertilizer prices on crop yields. Agricultural Economics. 2015;**46**(5):653-666

[10] Liadze I et al. Economic costs of the Russia-Ukraine war. The World Economy. 2023;**46**(4):874-886

[11] Kirilenko A, Dronin N. Recent grain production boom in Russia in historical context. Climatic Change. 2022;**171**(3):22

[12] FAO. World Fertilizer Trends and Outlook to 2022. 2019. Available from: https://openknowledge.fao.org/server/api/core/bitstreams/42d5a668-f44c-4976-8540-8efdb0f4d17b/content [Accessed: January 18, 2025]

[13] Wyer KE et al. Ammonia emissions from agriculture and their contribution to fine particulate matter: A review of implications for human health. Journal of Environmental Management. 2022;**323**:116285

[14] Elias SA. Global change impacts on the biosphere. In: Reference Module in Earth Systems and Environmental Sciences. Lexington, MA, USA: Elsevier; 2015

[15] Ellis C. World Bank: Global Waste Generation could Increase 70% by 2050; 2018. Available from: https://www.wastedive.com/news/world-bank-global-waste-generation-2050/533031/ [Accessed: January 17, 2025]

[16] Hao X, Chang C. Does long-term heavy cattle manure application increase salinity of a clay loam soil in semi-arid southern Alberta? Agriculture, Ecosystems and Environment. 2003;**94**(1):89-103

[17] Anthony AA, Oluyemi OA, Thor AS. Organic fertilizers: Public

health intricacies. In: Marcelo LL, Sonia S, editors. Organic Fertilizers. London, UK, Rijeka: IntechOpen; 2016. p. 16

[18] Khan MN et al. Fertilizers and their contaminants in soils, surface and groundwater. In: Dellasala DA, Goldstein MI, editors. Encyclopedia of the Anthropocene. Oxford: Elsevier; 2018. pp. 225-240

[19] Segura C et al. Effects on soil of grassland management for pasture, hay and silage. In: Goss MJ, Oliver M, editors. Encyclopedia of Soils in the Environment. Second ed. Oxford: Academic Press; 2023. pp. 102-112

[20] Somerville PD et al. Biochar and compost equally improve urban soil physical and biological properties and tree growth, with no added benefit in combination. Science of the Total Environment. 2020;**706**:135736

[21] Rodgers H et al. Sustaining vulnerable agroecosystems with compost: Lasting benefits to soil health and carbon storage in semiarid winter wheat (Triticum aestivum, L.). Applied Soil Ecology. 2024;**204**:105716

[22] Farhidi F, Madani K, Crichton R. How the US economy and environment can both benefit from composting management. Environmental Health Insights. 2022;**16**:11786302221128454

[23] Muhammad S et al. Composting sugarcane filter mud with different sources differently benefits sweet maize. Agronomy. 2023;**13**(3):748-748-748

[24] Colombini G et al. Coarse microplastic accumulation patterns in agricultural soils during two decades of different urban composts application. Environmental Pollution. 2024;**363**:125076

[25] Suzuki Y et al. Crop contamination evaluation by antimicrobial-resistant bacteria via livestock waste compost-fertilized field soil. Journal of Hazardous Materials. 2024;**480**:135987

[26] Farsang A et al. Evaluating the effects of sewage sludge compost applications on the microbial activity, the nutrient and heavy metal content of a Chernozem soil in a field survey. Arabian Journal of Geosciences. 2020;**13**(19):982

[27] Malinowska E. The effect of compost made with sewage sludge on heavy metal content in soil and in lolium multiflorum lam. Journal of Ecological Engineering. 2016;**17**(3):106-112

[28] Chu S et al. Municipal sewage sludge compost promotes Mangifera persiciforma tree growth with no risk of heavy metal contamination of soil. Scientific Reports. 2017;7(1):13408

[29] Culas RJ, Anwar MR, Maraseni TN. A framework for evaluating benefits of organic fertilizer use in agriculture. Journal of Agriculture and Food Research. 2025;**19**:101576

[30] Qi S et al. Systemic review for the use of biochar to mitigate soil degradation. GCB Bioenergy. 2024;**16**(6):1-28

[31] Singh P et al. A review on biochar composites for soil remediation applications: Comprehensive solution to contemporary challenges. Journal of Environmental Chemical Engineering. 2023;**11**(5):110635

[32] Fidel RB et al. Characterization and quantification of biochar alkalinity. Chemosphere. 2017;**167**:367-373

[33] Novak JM et al. Designing relevant biochars as soil amendments using lignocellulosic-based and manure-based

feedstocks. Journal of Soils and Sediments. 2014;**14**(2):330-343

[34] Razzaghi F, Obour PB, Arthur E. Does biochar improve soil water retention? A systematic review and meta-analysis. Geoderma. 2020;**361**:114055

[35] Castellini M et al. Impact of biochar addition on the physical and hydraulic properties of a clay soil. Soil and Tillage Research. 2015;**154**:1-13

[36] Maitra SAO et al. Bioinoculants-Natural Biological Resources for Sustainable Plant Production. LID - 10.3390/microorganisms10010051 [doi] LID - 51. 2021(2076-2607 (Print))

[37] Wu Y. Chapter 9 - periphyton: A promising bio-organic fertilizer source in agricultural ecosystems. In: Wu Y, editor. Periphyton. Boston: Elsevier; 2017. pp. 225-249

[38] Jaber LR, Enkerli J. Effect of seed treatment duration on growth and colonization of Vicia faba by endophytic Beauveria bassiana and Metarhizium brunneum. Biological Control. 2016;**103**:187-195

[39] Mącik M, Gryta A, Frąc M. Chapter two - biofertilizers in agriculture: An overview on concepts, strategies and effects on soil microorganisms. Advances in Agronomy. 2020;**162**:31-87

[40] Timofeeva A, Galyamova M, Sedykh S. Prospects for using phosphate-solubilizing microorganisms as natural fertilizers in agriculture. Plants (2223-7747). 2022;**11**(16):2119-2141

[41] Dey G et al. Management of phosphorus in salinity-stressed agriculture for sustainable crop production by salt-tolerant phosphate-solubilizing bacteria—A review. Agronomy. 2021;**11**(8):1552-1629

[42] Jalal A et al. Regulatory mechanisms of plant growth-promoting rhizobacteria and plant nutrition against abiotic stresses in brassicaceae family. Life (Basel). 2023;**13**(1):211-235

[43] Dincă LC et al. Fertilization and soil microbial community: A review. Applied Sciences (2076-3417). 2022;**12**(3):1198-1217

[44] Demir H et al. Biofertilizers improve the plant growth, yield, and mineral concentration of lettuce and broccoli. Agronomy. 2023;**13**(8):2031-2042

Section 2

Organic Materials Used as Fertilizers

Chapter 2

Biochar in Sustainable Agriculture: Fertilizer from Natural Resources?

Mari Carmen López-Pérez, Irene Iliana Ramírez-Bustos, Karla Ivette Chy-young Peñuñuri, Julio Javier Gastelum Reyes, Miriam Mendivil Morales and Fabián Pérez-Labrada

Abstract

Continuous and excessive use of synthetic fertilizers in food production generates strong pressure on agroecosystems. Therefore, it is necessary to study and standardize alternatives that promote environmental health, such as biochar. Biochar (a blackish solid obtained from the pyrolysis of vegetable waste, food industry waste, animal waste, or water processing waste) is characterized by its high content of mineral carbon, high presence of functional groups, and richness of nutrients such as potassium (K^+). Its use in soils improves the organic matter content and cation exchange capacity (CEC), which promotes crop nutrition. In addition to describing biochar manufacturing processes, this chapter aims to emphasize its role as a fertilizer, its stimulating effects on the soil microbiome and enzymatic activity, and its positive effects on plant health. Understanding biochar as a biofertilizer can be considered an ecological strategy for food production. However, the agricultural sector must consider the limitations of its production and use as fertilizer.

Keywords: pyrolysis, functional groups, cation exchange capacity, carbonization, soil-plant interactions

1. Introduction

In recent years, the use of biochar (BioC) has been widely studied, and great benefits in agricultural productivity have been reported, which can be considered as a strategy for managing agroecosystem wastes [1–3]. The use of crop and animal residues (straw and manure, respectively) as inputs makes BioC a potentially attractive alternative for recycling agroecosystem resources, managing carbon, and promoting its reservoir as soil organic carbon (SOC) [4]. BioC processing is based on the principle of pyrolysis (oxygen-free thermal decomposition of biomass), this processing mechanism allows to manage agricultural waste and to obtain a highly stable "solid waste" rich in functional groups and aromatic compounds on the surface [5, 6]. BioC is mainly composed of plant residues (woody and non-woody), food industry residues, animal waste, and sediments from wastewater treatment. Common sources are coconut husks, rice husks, wood chips, sawdust, corn, wheat, sorghum stubble, and cattle manure. In general, to produce BioC, the selected raw material is dried and ground and then subjected to pyrolysis at temperatures between 300 and 1000°C. In this process, the cellulose, hemicellulose, and lignin in the material undergo a series of changes, generating gaseous, liquid, and solid residues (BioC).

It is important to note that the type of material used, as well as the pyrolysis conditions, determine the main characteristics of BioC, such as high C, H, O, N, K, and S content, high porosity, high contact surface, functional groups (carbonyl, aliphatic, aromatic, hydroxyl), meso, macro, and micropores. Such characteristics can improve the water and nutrient-holding capacity, in addition to creating potential niches for microorganisms [7].

Thus, the use of BioC represents a sustainable method that allows the return of biomass to the agroecosystem and, owing to its characteristics, can help reduce chemical fertilizer rates [8]. Considering that this carbonaceous material is highly recalcitrant and persistent, it can improve some soil characteristics, such as microbiological activity [1], immobilize C in the microbial biomass [9], improve soil fertility and health, and increase crop yield and quality [10]. Similarly, a modification in the food web of microorganisms and soil health has been documented with the application of BioC [11], positively impacting sustainable agricultural production [12, 13]. In this sense, in some soil types, BioC can modulate the pH, water-holding capacity, and ion exchange potential of the soil [8], which are of vital importance and relevance for improving soil quality [14, 15].

Problems, such as inadequate chemical fertilization practices and low soil fertility [1], can lead to reduced organic matter content, reduced indigenous nutrient levels, and reduced microbiome load and diversity [1], which together cause declines in crop yield and quality. This constant and growing problem in food production can be ameliorated by integrating chemical fertilizers and BioC. For example, the use of BioC to recover and concentrate ortho-P and N-NH_4^+ from swine manure has recently been studied, opening the possibility of developing fertilizers for use in sustainable agriculture [16]. Similarly, it has been reported that in an eight-year study, the inclusion of BioC in the soil stimulates the abundance of functional microorganisms associated with the C, N, P, and S cycles, C degradation, nitrification, nitrate reduction, organic mineralization of P, and the S cycle [17]. Biochar and organic fertilizer applications enhance soil microbial abundance and agroecosystem multifunctionality. Additionally, BioC improves crop yield and quality [18, 19].

This chapter documents how BioC acts as a fertilizer in crops and describes the mechanisms that stimulate the microbiological load and enzymatic activity of the soil.

2. Biochar as fertilizer

BioC itself is not considered a fertilizer because of its heterogeneity and the high variability in its ionic content; however, here we propose BioC as a potential nutrient amendment and promoter of soil and plant quality (**Figure 1**). For example, in pakchoi, the use of 3% BioC stimulates morphological (plant height, root length, and dry weight) and leaf quality parameters (soluble sugars, vitamin C, and cellulose) along with a reduction in leaf nitrates. This response may be due to improved soil enzyme activity and rhizospheric nutrient content. The rhizospheric microbiological load increased markedly with BioC, mainly with *Alphaproteobacteria*,

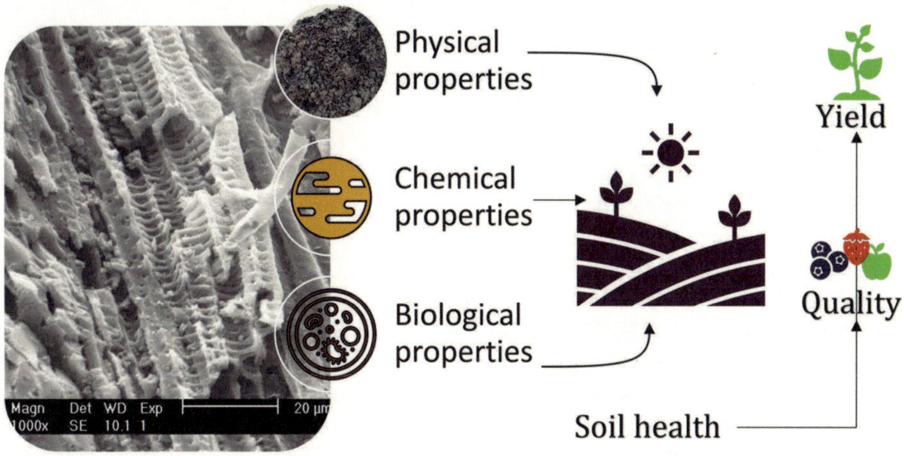

Figure 1.
Parameters potentially improved by the addition of biochar to the soil.

Gammaproteobacteria, *Bacteroidia*, *Acidimicrobiia*, *Illumatobacter*, *Luteolibacter*, *Lysobacter*, *Arthrobacter*, and *Mesorhizobium*. An increase in lipid transport and metabolism has also been documented [20].

Let us examine how BioC can promote certain soil parameters that fertilizers otherwise fail to enhance.

2.1 Stimulation of soil physicochemical parameters

Because BioC particles do not migrate in the soil profile in the short term [5], they can increase soil organic matter content due to their high C concentration [21], soil organic carbon (SOC) content, and cation exchange capacity (CEC) [22]. Similarly, BioC increased Haplic Luvisol soil pH and soil organic carbon and reduced NH_4^+-N and N_2O emissions [23]. In a semi-arid environment, wood BioC (>6.25 Mg ha^{-1}) can slightly increase soil pH and soil organic matter; however, it appears that this response is only short term (1st year of application) [24].

Surprisingly, BioC improved the soil profile macroaggregates (large and small) [25], size, distribution, proportion, and stability [26–28]. In this regard, BioC (25 Mg ha^{-1}) applied to Cambisol soils improves some hydrophysical properties of the soil, such as bulk density and porosity, probably by strengthening macropores and reducing hydraulic conductivity [29]. Owing to the characteristics of BioC, the pore size distribution is modified, increasing the proportion of micropores and improving water retention in the soil because these micropores have greater water retention owing to their capillary and adhesive forces [30].

Improved soil aggregates appear to lead to SOC conservation. However, this response seems to be regulated by the mean temperature, precipitation, C/N ratio, pH of BioC, BioC addition rate, fertilization rates, crop, and soil type [27]. It is reported that the use of BioC (at the rate of 3.2 Mg C ha^{-1}) slightly promotes soil quality index and ecosystem multifunctionality (capacity of soils to simultaneously provide multiple functions and services) derived from the improvement in aggregate diameter, nutrient content (C, N, and P), and enzyme activity associated with C, N and P acquisition [31].

2.2 Nutrient supply

It should be noted that the ionic charge of BioC varies according to the pyrolysis temperature and type of residue used. However, several studies have documented how this amendment provides ions such as K^+ and P to the soil [32]; for example, BioC (eucalyptus, 3.1 and 6.2 Mg ha^{-1}) promotes the availability and uptake of Ca^{2+} by rice plants in addition to improving the CEC, organic matter, available P, and exchangeable K^+. In sugarcane, it improved N, P, and K content [33], and in maize plants grown in tropical alkaline soils, the use of BioC (0.25, 0.5, and 1%, sugarcane filter cake, farmyard manure, and rice husk) promoted the availability of total N, nitrate (NO_3^--N), ammonium (NH_4^+-N), P-Olsen, extractable K, Zn, Mn, Cu, and Fe in the soil. In addition, an improvement in the fresh and dry biomass of the plant has been observed, as well as a higher concentration and absorption of nutrients (N, P, K, Mn, Cu, and Fe) [34].

When applying BioC from wood chips and poultry bones (pyrolyzed at 300 and 500°C) to Cambisol and Luvisol soil, the pH of BioC notably affected the mobility and bioaccessibility of Ca, Cu, K, Mg, Mn, P, S, and Zn in the medium term (360 days of incubation), indicating a slow but constant release of ions from BioC to the soil solution. However, it is necessary to clarify that this response is strongly related to BioC-soil interactions [7]. In another study, it was documented that the addition of BioC (10 t ha^{-1}) in addition to reducing salt and Na^+ content in the soil efficiently enhanced nutrients (total organic C, total N, NH_4^+-N, and NO_3^--N) and soil microbial biomass of carbon and nitrogen [35]. In the case of corn straw, BioC (10 t ha^{-1}) in soil cultivated with the rice-wheat rotation system promoted grain yield, probably because of an increase in total N content, available P, available K, and higher SOC content [4]. In the case of sheep manure, BioC (2 and 4%, pyrolyzed at 500°C) promoted NO_3^--N, Olsen-P, CEC, and NH_4^+-N contents in rice and tea soils. Likewise, it modified acid phosphomonoesterase and alkaline phosphomonoesterase activities. This type of BioC promotes P as an orthophosphate and an orthophosphate monoester [36].

Fertilization with BioC at a rate of 1.5% induced an increase in crop yield, which may have been due to improved sorption properties of the sandy soil and higher P, N, total carbon, and organic matter content [37]. The use of BioC (22 Mg ha^{-1}) promoted, after 2 years, crop water productivity and partial productivities of N, P, and K fertilization factors, improving soil organic C content in addition to reducing CH_4 and CO_2 emissions. However, the authors mentioned that there should be a BioC replenishment in the third year to replace C losses and improve soil C content. Similarly, the application of organic matter to soil is advised to optimize the positive responses of BioC [38]. Joint applications of BioC (3.5 Mg ha^{-1}) and straw (6 Mg ha^{-1}) improve soil organic matter (SOM) content, available nitrogen and potassium, dissolved organic carbon and particulate organic carbon, which is associated with the promotion of high microbial diversity and composition, improving paddy soil fertility [39].

The use of an Ultisol soil amended with BioC (5 t ha^{-1}) promoted okra yield in the greenhouse by improving soil pH, total N, organic carbon, available P, and exchangeable cations, leading to an improvement in height, number of leaves, leaf area, stem circumference, Ca, Mg, and K content in leaf tissue, as well as in the promotion of fruit yield [40]. As an organic fertilizer, BioC (6.0 Mg ha^{-1}) applied to Chernozem soil increased corn yield and plant growth rate, probably by increasing root volume, aerial biomass, and N uptake. Likewise, an improvement in soil quality was documented owing to the increase in total N, NH_4^+-N, enzyme activity associated with N acquisition (leucine aminopeptidase and alkaline phosphatase), and higher C associated with microbial biomass. Particularly, BioC enhanced the abundance of *Actinobacteria*, *Mortierellomycota*, *Chloroflexi*, *Acidobacteriota,* and *Ascomycota*, at different soil depths, activating soil nutrients [41].

Evidence suggests that another great advantage of BioC is that it minimizes nutrient leaching rates and nitrogen emissions [42, 43]. Therefore, in addition to directly adding C to the agroecosystem, BioC reduces the total reactive nitrogen content by minimizing its loss as nitrous and nitric oxide [44]. Some studies have shown that granular BioC promotes a lower leaching rate of N (up to 35%), K, Mg, and S, and a higher N release during the experiment. Interestingly, the granulated BioC maintained and recovered its microporosity. This new alternative facilitates its application in the field [45]. BioC from corn straw (10 t ha^{-1}) showed lower emissions of total methane (CH_4) and nitrous oxide (N_2O) in the short term [4].

One aspect of the vital relevance of BioC is that its effect on agricultural land varies with age; the initial application of BioC reduces N_2O emissions, but in the long term, no such effect is observed. This phenomenon may be due to fertilization practices (K^+ application rate), environmental interactions, soil characteristics (clay fraction), and the properties of the BioC used [46].

2.3 Fertilizer efficiency improvement

Providing 100% of a crop's nutrient requirements is technically and economically not feasible with the use of BioC alone [47]; therefore, this amendment can be used as an integral strategy to enhance the efficiency of chemical and organic fertilizer use.

2.3.1 BioC and chemical fertilizers

Encouraging results have been reported with the application of BioC-based fertilizer (0.6, 1.2, and 1.8 Mg ha^{-1}) in crops where nitrogen fertilization was reduced (up to 50%) because an increase in the content of easily oxidizable C, C-microbial, solid C, and soluble organic C in the soil was documented, as well as an increase in the net photosynthetic rate and stomatal conductance (first stages of the plant), transpiration rate, and CO_2-internal and dry biomass accumulation. Notably, this study applied a water regime via drip irrigation [48]. Similarly, the use of BioC in conjunction with nitrogen fertilization promotes soil microbial load: *Mesorhizobium*, *Porphyrobacter*, *Rhizobacter*, and *Allorhizobium-Neorhizobium-Parhizobium-Rhizobium* [49]. In a study after 4 years, BioC application succeeded in improving soil organic C, total N, and available P content in low-fertility rice soil [50], and interestingly, when applied in conjunction with chemical fertilizers (NPK), the responses were exacerbated.

In Haplic Luvisol soil cultivated with crop rotation (barley, corn, wheat, and peas), the application of BioC (10 and 20 t ha^{-1}) improved the efficiency of chemical fertilization, notably affecting the NO_3^--N, NH_4^+-N, and organic carbon content of the soil. However, the N mineralization rate depends on soil type and BioC [51]. Similarly, the combined application of nitrogen fertilization (277.78 kg ha^{-1}) and BioC (42.35 Mg ha^{-1}) improved the accumulation of SOC, dissolved organic carbon (DOC), and N fractions in the soil [52]. Effects of co-application of biochar and nitrogen. Similarly, BioC applications (15 and 30 t ha^{-1}) combined with nitrogen fertilization resulted in a reduction in soil pH, higher soil organic C content, mineral N, and higher P and K availability. This, in turn, improves the efficiency of N utilization in corn plants, which have high photosynthetic activity and greater accumulation of dry matter in leaves and grains [53].

In 3 years of rice and rapeseed rotation work, BioC (3, 9, 15, 15, 30, and 60 t ha^{-1}) coupled with NPK fertilizers (15-15-15) was found to promote SOC content, porosity, enzyme activity (urease, invertase, catalase, and alkaline phosphatase), nutrient uptake, and yield of rice and rapeseed [54]. The application of rice straw BioC to maize and wheat grown in an arid region promoted organic C content, available N, and CEC of clay soil without increasing salinity, and the soil pH was buffered by the alkaline nature of BioC. The combination with chemical fertilizers promotes the soluble and microbial fractions of organic carbon and nitrogen in the soil [47]. BioC also stimulates soil urease and dehydrogenase activity, thereby improving soil fertility.

In contrast, BioC applications (30 Mg ha^{-1}) in conjunction with fertilization ($N-P_2O_5-K_2O$) in tomatoes promoted SPAD units, soluble solids content, sugar-acid ratio, nutrient uptake, and dry matter yield. These benefits appear to improve under conditions of reduced irrigation [12], whereas in maize, the application of apple BioC (600°C; 3 Mg ha^{-1}) in combination with recommended fertilization promoted growth and yield [10], while a 30 Mg ha^{-1} dose of BioC increased water productivity, nitrogen use efficiency, and hence productivity [55]. In pakchoi, the combined use of BioC (0.5, 1, and 2% w/w) with nitrogen fertilization (100 and 200 mg N kg^{-1}) promotes yield in addition to improving soil microbiological load, particularly *Proteobacteria*, *Actinobacteriota*, *Gemmatimonadota*, *Acidobacteriota*, and *Chloroflexi* bacteria, and fungal communities *Ascomycota*, *Chytridiomycota*, *Mortierellomycota*, and *Basidiomycota* [56]. For maize plants grown in semi-arid regions, BioC application (20 Mg ha^{-1}) coupled with nitrogen fertilization promoted grain yield (11.1%) and reduced N_2O emissions [57].

2.3.2 BioC and organic fertilizers

Several studies corroborate the use of BioC in combination with organic sources (compost, vermicompost, manure, etc.). In this regard, it has been cited that BioC mixed with compost can supply K and P, improving soil C content [58]. BioC from seed layers in combination with organic fertilizers (flours derived from Camellia oleifera) promoted plant morphological parameters, biomass accumulation, and N, P, and K content in *Cunninghamia lanceolata* (Lamb.). In the acid soil-red, it promoted fertility (increasing organic matter, total N, NO_3^--N, total P, available P, total P, total K, and available K, and reducing NH_4^+-N). This practice differentially modifies soil bacterial and fungal (↑ *Gammaproteobacteria*, *Alphaproteobacteria*, *Leotiomycetes*, and *Sordariomycetes*; ↓ *Acidobacteriae* and *Agaricomycetes*) loads and biodiversity [17].

In 25-year-old olive trees, the use of compost BioC (90:10; % dw:dw) improved some physical properties of the soil (0–5 cm of the upper profile), such as bulk density, total porosity, aeration, and water-holding capacity. The mixture of BioC and compost promoted microbial activity and increased the content of organic-soluble C, N, and P, which helped generate aggregates [30]. The joint application of BioC from distillers' grains (1.5 and 3.0 Mg ha^{-1}) and vermicompost showed beneficial results, as higher vitamin C content and lower nitrate concentration in the fruit were found in chili peppers. It has also been reported that higher nutrient extraction as well as higher efficiency and recovery [59].

In dry no-tillage paddy soils, the application of BioC promoted the distribution and composition of soil bacteria and fungi, whereas combining BioC and organic fertilization resulted in the most complex association between soil microbial diversity, richness, and microbial ecological functions [14]. The combination of BioC and organic fertilizer (0.66 + 4.46 Mg ha^{-1}) evidenced a promotion in fertility, bacterial community composition, and an increase in bacterial diversity (*Proteobacteria*, *Actinobacteria*, *Gemmatimonadetes*, *Chloroflexi*, and *Bacteroidetes*) in the rhizospheric soil of potato crops. Higher concentrations of available P and soil organic carbon (SOC) have also been reported [60]. In 2-year field work, the addition of BioC (rice straw biochar, 15 t ha^{-1}) improved the nutrient availability of the organic fertilizer canola meal and promoted soil organic carbon, as well as the activity of the soil enzymes catalase, invertase, urease, and phosphatase [61]. The application of biochar improved the nutrient supply efficiency of the organic fertilizers. Microgranulated BioC (2%) supplemented with arbuscular mycorrhizal fungi improved the availability of extractable K^+ and Ca^{2+}, in addition to promoting the mycorrhization rate, root biomass, and P content, at least in peas [62]. BioC applications (alone, combined with manure, or enriched with urine + manure) in a company with recommended NPK fertilization on low-fertility soils positively promote soil pH, organic carbon, and soil N, P, and K content. These results suggest that BioC improves the efficiency of chemical fertilizer use [63].

However, the combined use of BioC, organic fertilizers, and chemical fertilizers can enhance the benefits of these fertilizers to a greater extent. For example, the basal application of BioC from swine manure (12 Mg ha^{-1}) in flooded rice fields together with standard fertilization (N-P_2O_5-K_2O = 90-45-57 kg ha^{-1}) reduced the volatilization of ammonia (NH_3) but did not promote grain productivity [64], which may be due to the low N content of BioC and its electronegative charge. In another study, it was found that the use of 16–23 Mg ha^{-1} of BioC (from agricultural residues and composted fecal amendments) in intensive systems (cattle manure + urea) increased soil organic C content (51%), whereas mineral fertilization decreased this variable by up to 32% after 6 months. The application per cycle of BioC promotes yield and alkalinizes the pH (from 6.5 to 7.4), although the positive response in the generation of fresh biomass depends on the crop. In cabbage, this positive response is appreciated, whereas in amaranth, this response was not observed [65].

Finally, the combined application of poultry litter BioC (20 Mg ha^{-1}) and chemical fertilizers improved soil nutrient status as well as SOC, soil microbial biomass carbon, nitrogen, and phosphorus contents. Surprisingly, BioC + chemical fertilization stimulated microbial abundance, mainly of actinomycetes, bacteria, mycorrhizae, and saprophytic fungi. Similarly, urease, dehydrogenase, and β-glucosidase enzyme activities were enhanced. All these factors improve wheat yield [66]. In another study, the use of BioC alone, combined with compost and/or NPK fertilization, improved pineapple growth and yield [67], whereas rice husk BioC (25 t ha^{-1}) increased the root yield of radish plants, but when applied in conjunction with chicken manure and chemical fertilization (NPK), improved nitrogen use efficiency, fertilizer recovery efficiency, agronomic efficiency, nitrogen harvest rate, and soil NH_4^+-N retention [68].

2.4 Soil microbiome activation

One of the characteristics that distinguishes BioC from fertilizers is its capacity to activate and improve the soil microbiological load and diversity [26, 69]. In this regard, the application of BioC improves C sequestration in soils by activating N-fixing bacteria, which can be explained by its high surface load and large number of pores [70], which can lead to a reduction in the nitrogen fertilization rate. Interestingly, BioC increases microbial respiration by up to 100% and seems to release more CO_2 from the soil, which is of vital importance because microorganisms degrade organic matter so that BioC can be decomposed slowly [5]. Thus, continuous application of low-dose BioC increases plant yields, partly due to the regulation of microbiological nitrification processes, that is, increased relative abundance of ammonia-oxidizing bacteria and a lower rate of nitrification by the reduction of *Nitrosomonas* and *Nitrosospira* [71]. Nine years of low-dose biochar amendment suppressed the nitrification rate in low-yield brown soils.

The use of BioC on *Jasminum sambac* enhanced the diversity of soil bacteria (*Lactobacillus*, *Azospirillum*, and *Cutibacterium*), promoting soil-associated organic carbon and nutrient decomposition through the activation of β-N-acetyl-glucosaminidase, β-1, 4-Glucosidase and β-D-cellobiosidase enzymes, that is, BioC is a source of carbonic fertilization [72], whereas in saline-alkaline soils (pH 10.31), BioC stimulates the composition and diversity of Pi bacteria (*Proteobacteria*, *Gammaproteobacteria*, *Alphaproteobacteria*, *Pseudomonas*, and *Rhizobium*), attenuating soil pH and salinity, allowing higher available P content [6]. Similarly, BioC alters the abundance of *Actinobacteria* and *Firmicutes* in saline soils [73], enhancing crop yield and microbial diversity. In some circumstances, BioC application at high doses (5%, peanut hulls, and straw) can cause negative alterations in the diversity and stability of some bacterial and archaeal communities, which in turn can modify the metabolism and nutrient cycling mediated by these microorganisms [74]. It is recommended to conduct studies aimed at discerning these contradictions.

2.5 Soil enzyme activity stimulation

Another great advantage of BioC as a fertilizer is its capacity to stimulate soil enzyme activity [75]; for example, BioC from corn stalks (40 Mg ha^{-1}) improves sucrase and catalase activity [32]. In the case of BioC applications (sunflower hulls and fruit treatment residues, at 650°C), enhanced and promoted enzyme activity was associated with C, N, and P, which was also associated with higher microbial respiration. Surprisingly, N, P, and K concentrations in soil pore water showed greater stability with BioC, indicating a reduction in leaching [76].

The use of BioC on Phaeozem and Luvisol soils differentially increased the content of organic C, organic P, inorganic forms of P, pyrophosphate, and orthophosphate. An increase in acid phosphomonoesterase, alkaline phosphomonoesterase, inorganic pyrophosphatase, and phosphodiesterase activities has also been documented. These results indicated that BioC promotes soil-dependent organic P turnover [77]. In contrast, the enzyme activities of β-glucosidase, acid and alkaline phosphatase, arylsulfatase, dehydrogenase, arylamidase, cellulase, and urease in the soil were promoted by the application of BioC (*Panicum virgatum*, 400–600°C), which is strongly associated with pH, organic matter, and CEC [78]. The higher activity of these enzymes indicates improved cellulose and carbohydrate decomposition, biological oxidation of organic matter, decomposition of nitrogenous organic compounds, release of inorganic phosphorus from the decomposition of organic forms, hydrolysis of urea (releasing NH_4^+ and CO_2), and release of sugars (glucose), which can potentially act as food for microorganisms.

In the case of manure, BioC promoted 22% leucine aminopeptidase enzyme activity compared to manure application, indicating that BioC limited N-associated microbial load [79]. When BioC was applied, a higher abundance of bacteria related to nitrate reduction, as well as cellulase and protease enzyme activities, was documented [80].

2.6 Stress mitigation

Another aspect of BioC as a fertilizer is the mitigation of stress in plants; for example, in soils with low fertility, the combined application of straw and BioC (derived from straw) promotes a higher load of gram-positive bacteria and fungi, which promotes the mineralization of soil organic matter. Thus, BioC, by participating as an electron carrier, promoting iron reduction and releasing organic carbon from soil minerals increasing soil organic matter mineralization [81]. In P-limited soils, the application of BioC (poultry litter biochar, 2%) showed a positive response by improving acid and alkaline phosphatase activity, phosphorus-utilizing bacterial load, and mycorrhizal colonization. Similarly, higher P uptake was observed in *Hordeum vulgare* plants. Accordingly, a higher relative abundance of *Actinobacteriota* and *Chloroflexi* has been observed [82]. In soils of semi-arid regions, the use of straw BioC increases the total and alkaline-hydrolyzable N content of the soil and improves the rate of N transport to the corn plant, improving the efficiency of water use and nitrogen fertilizer recovery [83].

In the case of acidic soils, rice, and sawdust BioC (3 and 12 Mg ha^{-1}) favored pumpkin yield owing to an increase in the organic carbon content of the soil, as well as greater efficiency and absorption of nutrients (N, P, K, Ca, and Mg). Interestingly, fruit quality (soluble sugar, soluble protein, and sugar-acid ratio) shows a positive response to BioC [84]. Likewise, the application of BioC (from pruned tea leaves) to soils in tea plantations resulted in an increase in soil pH as well as in the exchangeable ions K^+, Na^+, Ca^{2+}, and Mg^{2+}, and a reduction in the total exchangeable acidity and the concentration of exchangeable Al^{3+} and H^+ [85]. These results suggest an alternative method to stabilize the soil pH in tea plantations. In this type of soil, the use of BioC increases the pH, electrical conductivity, organic matter, phosphorus, nitrogen, and potassium available [86]. Rice husk BioC (15 and 20 Mg ha^{-1}) in soils irrigated with wastewater increased the soil pH, electrical conductivity, bulk density, SOC, total nitrogen, available phosphorus, enzyme activity, and soil microbial biomass. Interestingly, there is an increased microbial load of *Actinobacteria*, *Acidobacteria*, and *Nitrospira* [87].

The use of BioC (2 and 5%) on *Hordeum vulgare* L. plants grown under salinity stress (>50 mM) combined with NPK fertilizers significantly improved physiological (gas exchange, photosynthetic pigments, and SPAD value), biochemical (proline, soluble sugars, and phenolic compounds), mineral content, and grain yield [88]. Similarly, BioC of sugarcane bagasse (2%) restored the impact caused by soil salinity and promoted morphological parameters (shoot and root length), physiological parameters (relative water, protein, and carotenoid content), and grain yield of maize, in addition to reducing the Na^+ content in the roots and shoots. This type of BioC enhances the activity of urease and phosphatase (alkaline and acidic) enzymes [89].

In the case of salinity-alkalinity stress in maize plants, it was found that substituting one-third of the nitrogen fertilizer with BioC from cow dung (200 kg N ha^{-1} and BioC 100 kg N ha^{-1}), in addition to reducing CO_2 and N_2O emissions, improved yield, SOC storage, net ecosystem carbon budget, and agronomic N use efficiency [90]. In saline-alkaline soils (pH of 8.54, electrical conductivity of 6061 μS cm^{-1}), BioC applications (2%, rice straw) reduced pH, electrical conductivity, and water-soluble ions and increased soil organic matter, nutrient availability, and extracellular enzyme activity. Surprisingly, lower soil salinity was found, as well as a reduction in salts in cotton plant tissues. Similarly, there was a promotion of the growth and metabolism of salt-tolerant bacteria (*Proteobacteria*, *Bacteroidota*, *Acidobacteriota*, and *Actinobacteriota*), suggesting that BioC alleviated the salt damage to microorganisms by modifying soil metabolic pathways [91]. Alterations in the composition and metabolite profiles of saline-alkali soils.

In saline-alkaline soils, the use of BioC (2%) improved cotton plant biomass and boll dry weight, which could be due to an increase in photosynthetic pigment, carbohydrate, glutathione, and ascorbic acid content. It seems that BioC, in addition to reducing Na^+ uptake, improves the K^+/Na^+ and Ca^{2+}/Na^+ ionic ratios. In addition to the activation of the L-phenylalanine, piperidine, L-tryptophan, and allylsin biosynthesis pathways [92]. In highly saline-alkaline rice soils, BioC applications (1.5, 3.0, and 4.5%, w/w) combined with nitrogen fertilization (225 kg ha^{-1}) reduced pH, exchangeable sodium percentage, saturated paste extract, and sodium adsorption ratio, accompanied by an increase in CEC reflected in higher availability of P and K, as well as total and hydrolyzable nitrogen. The data also showed that BioC induced a higher abundance and structure of *Proteobacteria* [93].

Similarly, the use of BioC seems to reduce the availability of some metalloids (Cd, Cr, Ni, Zn, and Cu) by more than 50%, thereby minimizing the level of these elements in plants [87]. Joint applications of BioC and nanohydroxyapatite increased soil C, N, and P while reducing the bioavailability of Co, Cu, Fe, and Pb in the rhizosphere, which was expressed with a lower accumulation in soybean plants, possibly because BioC stimulates and restores the microbiological load of the soil [94]. Similarly, increasing the pH of the contaminated soil allowed the release of nutrients and stimulated microbiological activity. Similarly, BioC applications allow the conversion of acid-soluble forms to relatively inert reducible and oxidizable forms, thus minimizing Pb uptake. These data suggest a Pb phytostabilization capacity [95].

In *Panax quinquefolium* L., the use of BioC reduces Cd damage derived from (i) modifications in soil pH and organic matter content, (ii) promotion of beneficial soil microbial load (*Proteobacteria*, *Bacteroidota* and *Actinobacteria*), (iii) reduction of Cd content by soil chelation (mediated by naringenin, caffeic acid, and valine) and increase of detoxifying substances in the plant (malic acid, flavonoids, and fumaric acid) [96]. In *Sophora tonkinensis* plants grown on Cd-contaminated soil, the application of BioC (carbonized rice straw) and organic fertilizer (1:2) promoted a decrease in the total Cd content in the rhizosphere, coupled with increased nutrient and enzymatic activity of the soil. Similarly, the α diversity of bacterial load (*Acidobacteria*, *Proteobacteria*, *Bacteroidetes*, *Firmicutes*, *Chloroflexi*, *Gemmatimonadetes*, *Patescibacteria*, *Armatimonadetes*, Subgroup_6, *Bacillus*, and *Bacillus_Acidiceler*) was substantially enhanced, contributing to lower Cd content and higher accumulation of matrin and oxymatrine in plants [97]. Finally, the application of BioC (peanut, rice, maize, and wheat straw) to calcareous soils contaminated with Cd increased CEC, potentially improving soil fertility and reducing Cd availability [98].

As a strategy against drought, the use of BioC (wheat and corn straw) combined with water retention agents (polyacrylamide and sodium acrylate grafted with starch) is interesting because it increases field capacity, reduces volatilization and slows water infiltration due to an increase in the relative content of hydrophilic functional groups of Fluvisols soil (sandy loam and clay loam). Therefore, plants can improve their physiological resistance to drought by increasing K, soluble sugars, antioxidant enzyme activity, and reducing malondialdehyde content [99].

3. Biochar management strategies

This amendment can present certain rejection for its use in pristine form (bulk) and to a greater extent in soils with an alkaline reaction or high pH; therefore, several strategies have been documented to "condition" and enhance the properties of BioC (**Table 1**).

3.1 Accompanying

BioC derived from sustainable forestry logs, vinasse, and sugarcane molasses (500°C) increased the yield and palmitic acid content of *Cynara cardunculus* var. *Scolymus* is grown in a semi-arid region. This response can be explained by an increase in Ca, Mg, S, and Mn content in artichoke heads [100].

3.2 Acidification

Because of its alkaline nature, BioC can aggravate soil salinization problems; therefore, the use of acid-modified BioC (nitric and phosphoric) in addition to reducing the pH and electrical conductivity of the soil reduces the content of Na^+, Cl^-, SO_4^{2-}, CO_3^{2-},

Strategy	Results	Reference
BioC from sustainable forestry logs + vinasse + sugar cane molasses.	↑ Yield and palmitic acid content.	[100]
BioC modified with P and Mg.	↑ Organic C reserves, alteration in CO_2 release.	[101]
BioC loaded with Fe_3O_4 nanoparticles	↑ Cost-effective fertilization.	[8]
Sulfur-modified sunflower BioC (15 Mg ha^{-1})	↑ Microbial biomass carbon, basal respiration, enzyme activity, soil-dissolved organic carbon content, available nitrogen and phosphorus, silver N, P, and K content, higher yield. ↓ EC and Na^+ adsorption rate.	[102]
Iron-modified BioC	↑ N-NO3-, dissolved N- organic and N-total in soil. Enzyme activity β-glucosidase, β- Cellobiohydrolase and Leucine aminopeptidase. Community *Achromobacte, Gemmatimonas,* and *Cyanobacteriales*.	[103]
BioC modified with nitric and phosphoric acid.	↑ Enzymatic activity associated with C, N, and P. ↓ pH, EC, Na^+, Cl-, SO42-, CO32- and HCO3- content.	[104]
BioC produced by degassing with 0.1 M HCl and impregnated with urea.	↑ Prolonged N release. ↓ Volatilization of NH3, leaching of NO3- and emission of N2O.	[3]

Table 1.
Biochar management strategies as a source of fertilization in soils.

and HCO_3^- and stimulates the enzymatic activity associated with the acquisition of N, C, and P [104]. Similarly, BioC from *Acacia arabica* L. wood (produced by degassing with 0.1 M HCl) impregnated with urea showed prolonged N release, decreased 38% NH_3 volatilization, and minimized NO_3^- leaching by almost 50% and N_2O emission in alkaline-alkali calcareous soils [3]. Under greenhouse conditions, the combined use of fertilizers and modified BioC (physical, magnetic, biological, or acid), in addition to improving soil organic matter content and fertility, reduces greenhouse gas emissions [105]. Therefore, the acidification of BioC could be an alternative for use in alkaline soils [106].

3.3 Functionalization

Another alternative is functionalization, which consists of improving or increasing the ions or functional groups by adsorbing and complexing them on the BioC surface owing to its large surface area, porosity, and functional groups [3]. For example, some studies have reported that Fe-amended BioC alters the functional groups of BioC, and when applied to Luvisol soils (0.5 and 1%), it improves the NO_3^--N, dissolved organic N, and total N content and increases N accumulation in cotton plants. In addition to increasing the enzyme activity (β-glucosidase, β-cellobiohydrolase, and leucine aminopeptidase) and load on the soil microbial community (*Achromobacte*, *Gemmatimonas*, and *Cyanobacteriales*) [103]. BioC from banana peels loaded with Fe_3O_4 nanoparticles showed higher adsorption capacity than soil, proving to be a cost-effective and environmentally friendly fertilizer for *Abelmoschus esculentus* (Okra) [8]. In the cultivation of sunflower millet, BioC modified with sulfur (15 Mg ha^{-1}), water content, microbial biomass carbon, basal respiration, and enzymatic activity (catalase, dehydrogenase, urease, and alkaline phosphatase) improved, in addition to decreasing the electrical conductivity and Na^+ adsorption rate. Similarly, higher soil-dissolved organic C, N, and P-available contents have been reported, which led to an improvement in the N, P, and K content of silver [102].

The use of ions to be fixed in BioC alters its characteristics. In this respect, BioC (peanut shell, pyrolyzed at 500°C) modified with Mg ($MgCl_2$; 2 mol L^{-1}) minimized the release of CO_2 and N_2O, while modification with P (H_3PO_4, 50%) improved the release of CO_2 and decreased the release of N_2O. This may be due to the pH modifications that affect the metabolic activity of microorganisms associated with N_2 production. Overall, BioC+Mg is assumed to mitigate N emissions, allowing C reserves in coastal saline soils [101]. This response improves soil aggregates as the "charged" Mg in BioC allows binding between BioC+Mg with ions present in soil clay [107, 108].

Applications of biochar-coated nitrogen fertilizer in *Brassica rapa* subsp. Chinensis showed a promotion in chlorophyll content and fresh weight, in addition to a reduction in leaf nitrate content. The authors pointed out that this modified fertilizer seems to improve soil organic matter content in addition to minimizing the leaching of nitrate nitrogen, Ca^{2+}, and exchangeable Mg^{2+}. Together, these data indicate improved nitrogen use efficiency [109]. On the other hand, BioC enriched with nutrients (N, P, K) is promising in *Brassica napus* L. plants (1 and 5%, w/w), as it is documented to increase P and K availability, increase dehydrogenase enzyme activity, and induce higher nutrient concentrations in canola leaf tissue [110].

DOI: http://dx.doi.org/10.5772/intechopen.1008759

4. Plant health

BioC appears promising because it induces plant resistance, absorbs allelopathic and fungitoxic compounds, promotes beneficial microorganisms, and induces nutrient availability by improving soil health [111]. Thus, in addition to being a potential fertilizer, BioC can positively alter the microbial food web and soil health [11]. Thus, BioC seems to affect microorganisms because of its porous structure and high surface area (allowing it to be a sanctuary). As a source of C and N, it also seems to improve survival conditions by modifying soil pH, CEC, aggregation, and enzyme activity, in addition to immobilizing and generating greater nutrient residence owing to binding with labile soil C [112]. For example, it has been reported that rice straw BioC (25 Mg ha^{-1}) in *Brassica oleracea* L. var. italica Planch and *Brassica chinensis* L. promotes the degradation of pesticides (chlorantraniliprole, haloxyfop-ethotyl, and indoxacarb) and improves microbiological richness and diversity. This result depends on the type of pesticide, crop, and BioC used [113].

Similarly, the addition of BioC reduces insect herbivory and promotes mycorrhizal interactions in corn [114]. BioC also has a strong impact at the ecological level, and it has been documented that the use of this amendment seems to alter pH, water content, and soil surface temperature after 48 weeks, increasing the abundance of native ants (*Cardiocondyla nuda* and *Formica japonica*) and reducing the invasive ant species *Solenopsis invicta* [115].

Surprisingly, it appears that the dose of BioC plays a major role in soil microbial diversity and its primary components because, in addition to increasing total C and N in the soil, it promotes soil bacterial richness and uniformity by altering bacterial beta diversity (at doses >2% w/w). In particular, BioC positively affects *Acidobacteria*, *Bacteroidetes*, *Gemmatimonadetes*, and *Proteobacteria* (especially *Deltaproteobacteria* and *Gammaproteobacteria*), while *Verrucomicrobia* and *Basidiomycota* showed negative effects [116]. Similarly, the application of coconut shell BioC (1, 2, and 5 w/w) in tropical soils (latosol) differentially improved some physicochemical (pH and CEC), enzymatic (urease), and microbiological diversity parameters (*Proteobacteria*, *Gemmatimonadetes*, *Planctomycetes*, *Verrucomicrobia*, and *Bacteroidetes*), consequently reducing fungal strains, such as Ascomycota and some genera of *Haematonectria*, *Chaetomium*, *Gibberella*, *Aspergillus*, *Fusarium*, and *Eupenicillium* [117].

On the other hand, the combined application of BioC (rice husk, 2%) and humic acid (1.5%) reduced *Fusarium* abundance, helped recover soil microbial load in replanted apple orchards, and promoted soil urease, invertase, neutral phosphatase, and catalase activity. Together, these data improve soil health, leading to better fruit quality [118]. These responses may be because BioC can improve some physical (increasing aggregates, reducing bulk density, promoting permeability and water retention), chemical (improving N, P, and K use efficiency, increasing C reserves, reducing ammonia volatilization and N_2O and CO_2 emissions), and biological (activation of enzymes associated with absorption and transformation of organic matter, N, P, etc.) properties of the soil. The authors also pointed out that BioC can adsorb soil phenolic acids, which improves the plant resistance system [118].

Similarly, in apple trees, *Fusarium oxysporum* disease seems to be alleviated by the use of BioC (rice husk biochar, 2%) because this amendment reduces the abundance of this fungus in the soil, modifies the microbial load, and stimulates the edaphic environment by promoting the plant's root indices, in addition to potentiating the enzymatic activity of the soil (urease, invertase, neutral phosphatase, and catalase) and plant (superoxide dismutase, peroxidase, and catalase) [119]. Finally, as an amendment, 15 Mg ha^{-1} of BioC (palm residues pyrolyzed at 500°C) significantly improved available P and NH_4^+-N, in addition to promoting soil acid phosphatase, urease, and sucrase enzyme activity. Consistent with these results, BioC induced a higher relative abundance of *Rhodanobacteraceae*, *Stachybotryaceae*, *Agaricacea*, *Talaromyces*, *Nectriaceae*, *Sistotrema*, and *Bacillus*, accompanied by a decrease in *Fusarium* spp. (sugarcane Pokkah Boeng Disease Promoter) [120].

As a fertilizer, BioC presents a series of limitations that must be considered: (i) its production at the industrial level can be costly, reducing its profitability and convenience; (ii) the price can be high (USD \$5.00 kg^{-1}); (iii) generally, the recommended application doses are very high, increasing the initial cost; (iv) smaller-scale production can generate a reduced quality product; (v) inadequate pyrolysis control can release carbon monoxide and hydrocarbons into the environment; (vi) sometimes the BioC-biomass yield is reduced, requiring high amounts of raw material, (vii) the benefits can be medium or long term and (viii) the quality of the product varies greatly depending on the material and pyrolysis temperature used. Therefore, its use in sustainable agriculture should be considered.

5. Conclusion

As an amendment to the food production system, biochar seems to be a resource with high potential to improve the sustainability of agriculture, as it can be used to manage agroecosystem residues, and by acting as a carbon reservoir, it contributes a reserve of vital importance for plants. Likewise, its use has great benefits compared with chemical fertilizers, such as reducing biotic and abiotic stress in plants. Thus, biochar is an alternative for sustainable management of agroecosystems.

The use of biochar as a soil amendment can improve fertilizer efficiency, activate soil microbiological load and diversity, stimulate enzyme activity, and improve soil health and nutritional status. In addition to improving soil aggregates, it can also stimulate soil organic carbon content. It should be noted that the benefits of biochar depend strongly on the type of material, time, and type of pyrolysis, but also on the type of soil; therefore, it is necessary to understand these parameters at greater depths to standardize its use and implement it as another strategy in the sustainable production system of agroecosystems. In addition, the use of biochar in sustainable production must consider the limitations of its production and use as a fertilizer, which is why this carbonaceous material is presented as another alternative for the agricultural sector.

Acknowledgements

The authors are highly thankful to the researchers for their findings in this area, which served in the preparation of this manuscript.

Funding

The authors received no direct funding for this study.

Conflict of interest

The authors declare no conflicts of interest.

Data availability

All the data generated are included in this article reference's part.

Author details

Mari Carmen López-Pérez[1], Irene Iliana Ramírez-Bustos[2],
Karla Ivette Chy-young Peñuñuri[1], Julio Javier Gastelum Reyes[1],
Miriam Mendivil Morales[1] and Fabián Pérez-Labrada[3]*

1 Universidad Estatal de Sonora, Navojoa Sonora, México

2 Universidad Estatal de Sonora, Hermosillo Sonora, México

3 Universidad Autónoma Agraria Antonio Narro, Saltillo Coahuila, México

*Address all correspondence to: fabian.perezl@uaaan.edu.mx

References

[1] Imran. Integration of organic, inorganic and bio fertilizer, improve maize-wheat system productivity and soil nutrients [Internet]. Journal of Plant Nutrition. 2024;**47**:2494-2510. DOI: 10.1080/01904167.2024.2354190

[2] Marathe S, Sadowski Ł. Developments in biochar incorporated geopolymers and alkali activated materials: A systematic literature review. Journal of Cleaner Production. 2024;**469**:143136

[3] Rashid M, Hussain Q, Hayat R, Ahmad M, Azeem M, Alvi S, et al. Deashed biochar as N-carrier extended the N-release by inhibiting N-losses in calcareous soils. Biomass Conversion and Biorefinery. 2023;**13**(11):9549-9564. DOI: 10.1007/s13399-023-04250-5

[4] Xia X, Zhao Z, Ding Y, Feng X, Chen S, Shao Q, et al. Crop residue biochar rather than manure and straw return provided short term synergism among grain production, carbon sequestration, and greenhouse gas emission reduction in a paddy under rice-wheat rotation. Food and Energy Security. 2024;**13**(5):e70009. DOI: 10.1002/fes3.70009

[5] Galitskaya PY, Selivanovskaya SY, Karamova KO, Gordeev AS, Kuryntseva PA, Ghorbannezhad P. Transformation of biochar from plant biomass in gray forest soil: Evaluation by isotopic labeling method. Eurasian Soil Science. 2024;**57**(10):1579-1589. DOI: 10.1134/S1064229324601598

[6] Zhao W, Zhao H, Sun X, Wang H, Sun Y, Liang Y, et al. Biochar and wood vinegar altered the composition of inorganic phosphorus bacteria community in saline-alkali soils and promoted the bioavailability of phosphorus. Journal of Environmental Management. 2024;**370**:122501

[7] Száková J, Stiborová H, Mercl F, Hailegnaw NS, Lhotka M, Derevyankina T, et al. Woodchips biochar versus bone char in a one-year model soil incubation experiment: The importance of soil/char pH alteration on nutrient availability in soil. Journal of Chemical Technology and Biotechnology. 2023;**99**(10):2186-2197. DOI: 10.1002/jctb.7421

[8] Dar A, Hafeez M, Sarwar F, ul Ain N, Yaseen G. Iron-doped biochar, an agricultural and environmentally beneficial fertilizer. Environmental Monitoring and Assessment. 2024;**196**(6):1-11. DOI: 10.1007/s10661-024-12695-y

[9] Massaccesi L, Nogués I, Mazzurco Miritana V, Passatore L, Zacchini M, Pietrini F, et al. Short-term effects of biochar and compost on soil microbial community, C and N cycling, and lettuce (Lactuca sativa L.) yield in a Mediterranean environment. Applied Soil Ecology. 2024;**199**:105411

[10] Wani OA, Akhter F, Kumar SS, Kanth RH, Dar ZA, Babu S, et al. Pyrolyzed and unpyrolyzed residues enhance maize yield under varying rates of application and fertilization regimes. PeerJ. 2024;**12**(6):e17513. Available from: https://peerj.com/articles/17513

[11] Zhuang W, Zhao C, Zhang Y, Yang Z, Li G, Su L, et al. Synergistic application of biochar with organic fertilizer positively impacts the soil micro-food web in sandy loam soils. European Journal of Soil Biology. 2024;**123**:103680

[12] Zhao W, Wu K, Wu Y, Yu H, Cao W, Ma H. Effects of biochar amendment

on greenhouse tomato quality, nutrient uptake and use efficiency under various irrigation and fertilization regimes. Scientia Horticulturae. 2024;**337**:113441

[13] Zhang X, Feng X, Chai N, Kuzyakov Y, Zhang F, Li FM. Biochar effects on crop yield variability. Field Crops Research. 2024;**316**:109518

[14] Cao X, Liu J, Zhang L, Mao W, Li M, Wang H, et al. Response of soil microbial ecological functions and biological characteristics to organic fertilizer combined with biochar in dry direct-seeded paddy fields. Science of the Total Environment. 2024;**948**:174844

[15] Faloye OT, Ajayi AE, Kamchoom V, Akintola OA, Oguntunde PG. Evaluating impacts of biochar and inorganic fertilizer applications on soil quality and maize yield using principal component analysis. Agronomy. 2024;**14**(8):1761. Available from: https://www.mdpi.com/2073-4395/14/8/1761/htm

[16] Yadav K, Bakshi S, Banik C, Andersen DS, Brown RC. Assessing biochar and zeolite for enhanced agricultural sustainability of swine manure. Journal of Environmental Chemical Engineering. 2024;**12**(3):112987

[17] Hu Y, Wang Y, Tang Y, Yang H, Fan S, Zeng Y, et al. Biochar combined with organic fertilizer improved soil quality, promoted the growth of *Cunninghamia lanceolata* (Lamb.) hook and changed the microbial community. Ecoscience. 2024;**31**(3):149-164. DOI: 10.1080/11956860.2024.2406162

[18] Ali I, He L, Ullah S, Quan Z, Wei S, Iqbal A, et al. Biochar addition coupled with nitrogen fertilization impacts on soil quality, crop productivity, and nitrogen uptake under double-cropping system. Food and Energy Security. 2020;**9**(3):e208. DOI: 10.1002/fes3.208

[19] Amirahmadi E, Ghorbani M, Krexner T, Hörtenhuber SJ, Bernas J, Neugschwandtner RW, et al. Life cycle assessment of biochar and cattle manure application in sugar beet cultivation – Insights into root yields, white sugar quality, environmental aspects in field and factory phases. Journal of Cleaner Production. 2024;**476**:143772

[20] Wu X, Yang F, Zhang J, Gao F, Hu YC, Yang K, et al. Biochar's role in improving pakchoi quality and microbial community structure in rhizosphere soil. PeerJ. 2024;**12**:e16733. Available from: https://peerj.com/articles/16733

[21] Lopes JI, Arrobas M, Raimundo S, Gonçalves A, Brito C, Martins S, et al. Photosynthesis, yield, nutrient availability and soil properties after biochar, zeolites or mycorrhizal inoculum application to a mature rainfed olive orchard. Agriculture (Switzerland). 2022;**12**(2):171. Available from: https://www.mdpi.com/2077-0472/12/2/171/htm

[22] Scotti C, Bertora C, Valagussa M, Borrelli L, Cabassi G, Tosca A. Agroenvironmental performances of biochar application in the mineral and organic fertilization strategies of a maize– Ryegrass forage system. Agriculture (Switzerland). 2022;**12**(7):925. Available from: https://www.mdpi.com/2077-0472/12/7/925/htm

[23] Kotuš T, Šimanský V, Drgoňová K, Illéš M, Wójcik-Gront E, Balashov E, et al. Combination of biochar with N–fertilizer affects properties of soil and N_2O emissions in maize crop. Agronomy. 2022;**12**(6):1314. Available from: https://www.mdpi.com/2073-4395/12/6/1314/htm

[24] Blanco-Canqui H, Creech CF, Easterly AC. How does biochar

impact soils and crops in a semi-arid environment? A 5-yr assessment. Field Crops Research. 2024;**310**:109340

[25] Zhao H, Wang Z, Liu H, Xiao H, Gao M. Water-stable aggregates and aggregate-associated organic carbon after two years of biochar application. Archives of Agronomy and Soil Science. 2023;**69**(11):2218-2232. DOI: 10.1080/03650340.2022.2142937

[26] Bai N, Zhang H, Zhou S, Sun H, Zhao Y, Zheng X, et al. Long-term effects of straw return and straw-derived biochar amendment on bacterial communities in soil aggregates. Scientific Reports. 2020;**10**(1):1-10. Available from: https://www.nature.com/articles/s41598-020-64857-w

[27] Xu P, Wang Q, Duan C, Huang G, Dong K, Wang C. Biochar addition promotes soil organic carbon sequestration dominantly contributed by macro-aggregates in agricultural ecosystems of China. Journal of Environmental Management. 2024;**359**:121042

[28] Wu J, Teng B, Zhong Y, Duan X, Gong L, Guo W, et al. Enhancing soil aggregate stability and organic carbon in Northwestern China through straw, biochar, and nitrogen supplementation. Agronomy. 2024;**14**(5):899. Available from: https://www.mdpi.com/2073-4395/14/5/899/htm

[29] Ayaz M, Feizienė D, Feiza V, Tilvikienė V, Baltrėnaitė-Gedienė E, Khan A. The impact of swine manure biochar on the physical properties and microbial activity of loamy soils. Plants. 2022;**11**(13):1729. Available from: https://www.mdpi.com/2223-7747/11/13/1729/htm

[30] Fornes F, Lidón A, Belda RM, Macan GPF, Cayuela ML, Sánchez-García M, et al. Soil fertility and plant nutrition in an organic olive orchard after 5 years of amendment with compost, biochar or their blend. Scientific Reports. 2024;**14**(1):1-12. Available from: https://www.nature.com/articles/s41598-024-67565-x

[31] Wang C, Ma Y, He W, Kuzyakov Y, Bol R, Chen H, et al. Soil quality and ecosystem multifunctionality after 13-year of organic and nitrogen fertilization. Science of the Total Environment. 2024;**931**:172789

[32] Yan S, Zhang S, Yan P, Aurangzeib M. Effect of biochar application method and amount on the soil quality and maize yield in Mollisols of Northeast China. Biochar. 2022;**4**(1):1-15. DOI: 10.1007/s42773-022-00180-z

[33] Butphu S, Rasche F, Cadisch G, Kaewpradit W. Eucalyptus biochar application enhances Ca uptake of upland rice, soil available P, exchangeable K, yield, and N use efficiency of sugarcane in a crop rotation system. Journal of Plant Nutrition and Soil Science. 2020;**183**(1):58-68. DOI: 10.1002/jpln.201900171

[34] Choudhary TK, Khan KS, Hussain Q, Ashfaq M. Nutrient availability to maize crop (Zea mays L.) in biochar amended alkaline subtropical soil. Journal of Soil Science and Plant Nutrition. 2021;**21**(2):1293-1306. DOI: 10.1007/s42729-021-00440-0

[35] Wang S, Gao P, Zhang Q, Shi Y, Guo X, Lv Q, et al. Biochar improves soil quality and wheat yield in saline-alkali soils beyond organic fertilizer in a 3-year field trial. Environmental Science and Pollution Research. 2023;**30**(7):19097-19110. DOI: 10.1007/s11356-022-23499-3

[36] Hosseini SH, Liang X, Niyungeko C, Zakeritabar SF. Soil nutrients, phosphatase activities and phosphorus

species distribution in soils amended sheep manure biochar. Journal of Soils and Sediments. 2024;**24**(9):3297-3310. DOI: 10.1007/s11368-024-03869-9

[37] Stańczyk-Mazanek E. Analysis of the effect of sandy soil amendment with biochar on its physical and chemical properties and the quantity and quality of biomass yield of energy crops. Desalination and Water Treatment. 2023;**301**:216-227

[38] Jia Y, Xu Z, Zhang D, Yang W, Ding Y, Li C, et al. Effects of a single biochar application on soil carbon storage and water and fertilizer productivity of drip-irrigated, film-mulched maize production. Agronomy. 2024;**14**(9):2028. Available from: https://www.mdpi.com/2073-4395/14/9/2028/htm

[39] Xia H, Shen J, Riaz M, Jiang C, Zu C, Jiang C, et al. Effects of biochar and straw amendment on soil fertility and microbial communities in paddy soils. Plants. 2024;**13**(11):1478. Available from: https://www.mdpi.com/2223-7747/13/11/1478/htm

[40] Sangotoye SE, Atere CT, Taiwo LB, Olayinka A. Improving soil chemical properties and okra (*Abelmoschus esculentus*) yield through complementary applications of biochar, and organic and inorganic fertilizers. Journal of Plant Nutrition. 2024;**47**(15):2530-2542. DOI: 10.1080/01904167.2024.2354193

[41] Feng W, Sánchez-Rodríguez AR, Bilyera N, Wang J, Wang X, Han Y, et al. Mechanisms of biochar-based organic fertilizers enhancing maize yield on a Chinese Chernozem: Root traits, soil quality and soil microorganisms. Environmental Technology and Innovation. 2024;**36**:103756

[42] Huang Y, Tao B, Lal R, Lorenz K, Jacinthe PA, Shrestha RK, et al. A global synthesis of biochar's sustainability in climate-smart agriculture - evidence from field and laboratory experiments. Renewable and Sustainable Energy Reviews. 2023;**172**:113042

[43] Preza Fontes G, Greer KD, Pittelkow CM. Does biochar improve nitrogen use efficiency in maize? GCB Bioenergy. 2024;**16**(1):e13122. DOI: 10.1111/gcbb.13122

[44] Lin H, Guo S, Han Z, Liu S, Wang J, Zou J. Can biochar application improve the net economic benefits of tea plantations? Science of the Total Environment. 2023;**856**:159029

[45] Grafmüller J, Möllmer J, Muehe EM, Kammann CI, Kray D, Schmidt HP, et al. Granulation compared to co-application of biochar plus mineral fertilizer and its impacts on crop growth and nutrient leaching. Scientific Reports. 2024;**14**(1):1-15. Available from: https://www.nature.com/articles/s41598-024-66992-0

[46] Wang S, Li J, Yuan X, Senadheera SS, Chang SX, Wang X, et al. Machine learning predicts biochar aging effects on nitrous oxide emissions from agricultural soils. ACS Agricultural Science & Technology. 2024;**4**(9):888-898. DOI: 10.1021/acsagscitech.4c00114

[47] Mohamed I, El-habbak AK, Abbas MH, Scopa A, Drosos M, AbdelRahman MAE, et al. Rice straw biochar and NPK minerals for sustainable crop production in arid soils: A case study on maize-wheat cropping system. CABI Agriculture and Bioscience. 2024;**5**(1):1-16. DOI: 10.1186/s43170-024-00289-0

[48] Ren T, Fan P, Zuo W, Liao Z, Wang F, Wei Y, et al. Biochar-based fertilizer under drip irrigation: More conducive to improving soil carbon pool

and promoting nitrogen utilization. Ecological Indicators. 2023;**154**:110583

[49] Pang Z, Huang J, Fallah N, Lin W, Yuan Z, Hu C. Combining N fertilization with biochar affects root-shoot growth, rhizosphere soil properties and bacterial communities under sugarcane monocropping. Industrial Crops and Products. 2022;**182**:114899

[50] Chen S, Liu G, Hong Y, Ma Y, Guo S, Yan P, et al. Biochar amendment was less effective for Rice yield improvement but more effective for soil quality relative to inorganic fertilization in a low fertility Paddy soil. Journal of Soil Science and Plant Nutrition. 2024;**24**(3):4918-4928. DOI: 10.1007/s42729-024-01881-z

[51] Hematimatin N, Igaz D, Aydın E, Horák J. Biochar application regulating soil inorganic nitrogen and organic carbon content in cropland in the Central Europe: A seven-year field study. Biochar. 2024;**6**(1):1-17. DOI: 10.1007/s42773-024-00307-4

[52] Jia X, Ma H, Yan W, Shangguan Z, Zhong Y. Effects of co-application of biochar and nitrogen fertilizer on soil profile carbon and nitrogen stocks and their fractions in wheat field. Journal of Environmental Management. 2024;**368**:122140

[53] Li C, Zhao C, Zhao X, Wang Y, Lv X, Zhu X, et al. Beneficial effects of biochar application with nitrogen fertilizer on soil nitrogen retention, absorption and utilization in maize production. Agronomy. 2023;**13**(1):113. Available from: https://www.mdpi.com/2073-4395/13/1/113/htm

[54] Khan Z, Zhang K, Khan MN, Zhu K, Hu L. Effects of biochar persistence on soil physiochemical properties, enzymatic activities, nutrient utilization, and crop yield in a three-year rice-rapeseed crop rotation. European Journal of Agronomy. 2024;**154**:127096

[55] Zhang P, Liu J, Wang M, Zhang H, Yang N, Ma J, et al. Effects of irrigation and fertilization with biochar on the growth, yield, and water/nitrogen use of maize on the Guanzhong plain, China. Agricultural Water Management. 2024;**295**:108786

[56] Zhong L, Gu Z, Sun Y, Wang R, Wang H, Li G, et al. Biochar increases pakchoi yield by regulating soil bacterial communities but reduces it through soil fungi in vegetable soil. Journal of Soils and Sediments. 2024;**24**(3):1348-1360. DOI: 10.1007/s11368-024-03733-w

[57] Zhou J, Yue S, Liu Q, Zeng X, Song J, Bao S, et al. Impact of biochar application on nitrous oxide and methane emissions in rainfed cropping systems within a semiarid region. Agronomy Journal. 2024;**116**(6):3007-3020. DOI: 10.1002/agj2.21711

[58] Nobile C, Lebrun M, Védère C, Honvault N, Aubertin ML, Faucon MP, et al. Biochar and compost addition increases soil organic carbon content and substitutes P and K fertilizer in three French cropping systems. Agronomy for Sustainable Development. 2022;**42**(6):1-15. DOI: 10.1007/s13593-022-00848-7

[59] Zhang M, Liu Y, Wei Q, Liu L, Gu X, Gou J, et al. Effects of biochar and vermicompost on growth and economic benefits of continuous cropping pepper at karst yellow soil region in Southwest China. Frontiers in Plant Science. 2023;**14**:1238663

[60] Hou J, Xing CF, Zhang J, Wang Z, Liu M, Duan Y, et al. Increase in potato yield by the combined application of biochar and organic fertilizer: Key role of rhizosphere microbial diversity. Frontiers in Plant Science. 2024;**15**:1389864

[61] Zhang K, Khan Z, Khan MN, Luo T, Luo L, Bi J, et al. The application of biochar improves the nutrient supply efficiency of organic fertilizer, sustains soil quality and promotes sustainable crop production. Food and Energy Security. 2024;**13**(1):e520. DOI: 10.1002/fes3.520

[62] Fišarová L, Berchová K, Lukáč M, Beesley L, Vosátka M, Hausenblas M, et al. Microgranular biochar improves soil fertility and mycorrhization in crop systems. Soil Use and Management. 2024;**40**(2):e13068. DOI: 10.1111/sum.13068

[63] Pandit NR, Sipkhan P, Sharma SS, Dawadi D, Vista SP, Raut P. Cattle-urine-enriched biochar enhances soil fertility, nutrient uptake, and yield of maize in a low-productive soil. Nitrogen (Switzerland). 2024;**5**(1):16-27. Available from: https://www.mdpi.com/2504-3129/5/1/2/htm

[64] Canatoy RC, Cho SR, Galgo SJC, Park SY, Kim PJ. Biochar manure decreases ammonia volatilization loss and sustains crop productivity in rice paddy. Frontiers in Environmental Science. 2024;**12**:1421320

[65] Manka'abusi D, Zongo N, DJP L, Steiner C, Marschner B, Buerkert A. Soil properties and agronomic effects of repeated biochar amendment and fertilization in an urban horticultural system of Burkina Faso. European Journal of Agronomy. 2024;**159**: 127234

[66] Aziz MA, Khan KS, Khalid R, Shabaan M, Alghamdi AG, Alasmary Z, et al. Integrated application of biochar and chemical fertilizers improves wheat (Triticum aestivum) productivity by enhancing soil microbial activities. Plant and Soil. 2024;**502**(1):433-448. DOI: 10.1007/s11104-024-06556-3

[67] Hanyabui E, Frimpong KA, Annor-Frempong F, Atiah K. Effect of pineapple waste biochar and compost application on the growth and yield of pineapple varieties in Ghana. Frontiers in Agronomy. 2024;**6**:1331377. DOI: 10.3389/fagro.2024.1331377

[68] Mon WW, Ueno H. Short-term effect of the combined application of Rice husk biochar and organic and inorganic fertilizers on radish growth and nitrogen use efficiency. Plants. 2024;**13**(17):2376. Available from: https://www.mdpi.com/2223-7747/13/17/2376/htm

[69] Duan Y, Zhang L, Yang J, Zhang Z, Awasthi MK, Li H. Insight to bacteria community response of organic management in apple orchard-bagasse fertilizer combined with biochar. Chemosphere. 2022;**286**:131693

[70] Castejón-del Pino R, Cayuela ML, Sánchez-García M, Siles JA, Sánchez-Monedero MA. Field assessment of biochar interactions with chemical and biological N fertilization in pointed white cabbage. GCB Bioenergy. 2024;**16**(11):e70006. DOI: 10.1111/gcbb.70006

[71] Li Y, Wei H, Liu M, Yang J, Han X. Nine years of low-dose biochar amendment suppresses nitrification rate in low-yield brown soil. Applied Soil Ecology. 2023;**192**:105096

[72] Yang F, Wang W, Wu Z, Peng J, Xu H, Ge M, et al. Fertilizer reduction and biochar amendment promote soil mineral-associated organic carbon, bacterial activity, and enzyme activity in a jasmine garden in Southeast China. Science of the Total Environment. 2024;**954**:176300

[73] Zhao X, Grossart HP. Enhancing crop yield and microbial diversity in saline-affected paddy soil through

biochar amendment under aquaculture wastewater irrigation. European Journal of Soil Biology. 2024;**123**: 103681

[74] Zhu N, Yu Q, Song L, Sheng H. The inhibiting effects of high-dose biochar application on soil microbial metagenomics and Rice (Oryza sativa L.) production. International Journal of Molecular Sciences. 2023;**24**(20):15043. Available from: https://www.mdpi.com/1422-0067/24/20/15043/htm

[75] Antileo-Mellado S, Muñoz C, Sanchez-Hernandez JC, Ginebra M, Sandoval M. Effect of biochar on Vermicompost production: Chemical, biochemical, and biological properties. Agronomy. 2024;**14**(3):615. Available from: https://www.mdpi.com/2073-4395/14/3/615/htm

[76] Brtnicky M, Mustafa A, Hammerschmiedt T, Kintl A, Trakal L, Beesley L, et al. Pre-activated biochar by fertilizers mitigates nutrient leaching and stimulates soil microbial activity. Chemical and Biological Technologies in Agriculture. 2023;**10**(1):1-14. DOI: 10.1186/s40538-023-00430-7

[77] Liu X, Wu C, Jiang D, Zhang Y, Chen Z. Biochar application regulates organic phosphorus fractions and the release of available phosphorus in farmland soil. Journal of the Science of Food and Agriculture. 2024;**105**(1): 671-681. DOI: 10.1002/jsfa.13863

[78] Zaid F, Al-Awwal N, Yang J, Anderson SH, Alsunuse BTB. Effects of biochar-amended composts on selected enzyme activities in soils. PRO. 2024;**12**(8):1678. Available from: https://www.mdpi.com/2227-9717/12/8/1678/htm

[79] An Z, Gross CD, Chen X, Bork EW, Carlyle CN, Chang SX. Manure and its biochar affect activities and stoichiometry of soil extracellular enzymes in croplands. Journal of Soils and Sediments. 2024;**24**(9):3286-3296. DOI: 10.1007/s11368-024-03887-7

[80] Tang Z, He N, Zhang L, Wang L, Gong D, Wang C, et al. Straw and biochar application alters the structure of rhizosphere microbial communities in direct-seeded Rice (Oryza sativa L.) paddies. Agronomy. 2024;**14**(2):316. Available from: https://www.mdpi.com/2073-4395/14/2/316/htm

[81] Liu Q, Wu C, Wei L, Wang S, Deng Y, Ling W, et al. Microbial mechanisms of organic matter mineralization induced by straw in biochar-amended paddy soil. Biochar. 2024;**6**(1):1-13. DOI: 10.1007/s42773-024-00312-7

[82] Deinert L, Hossen S, Ikoyi I, Kwapinksi W, Noll M, Schmalenberger A. Poultry litter biochar soil amendment affects microbial community structures, promotes phosphorus cycling and growth of barley (Hordeum vulgare). European Journal of Soil Biology. 2024;**120**:103591

[83] Wang Y, Zhang M, Tian Z, Yang J, Pang J, Fang Y, et al. Incorporating straw-derived biochars enhances soil nitrogen accumulation and mitigates nitrogen leaching to facilitate crop productivity of mulching farmland in semiarid regions. Land Degradation and Development. 2024;**35**(15):4484-4496. DOI: 10.1002/ldr.5234

[84] Li Z, Hao Y, Li H, Fu T, Li J, Peng Y, et al. Biochar improves wax gourd (Benincasa hispida (Thunb.) Cogn.) yield and quality by regulating the chemical properties of acidic soil and promoting nutrient uptake. Agronomy. 2024;**14**(8):1750. Available from: https://www.mdpi.com/2073-4395/14/8/1750/htm

[85] Kelin C, Yiheng L, Heng Z, Kang L, Yunfei Y, Yingjie Z, et al. The effect of pruned tea leave biochar and cattle manure for soil acidification. Soil Use and Management. 2024;**40**(3):e13091. DOI: 10.1111/sum.13091

[86] Aurangzeib M, Zhang S, Yan S, Zhou J, Niu X, Yan P, et al. Biochar application can improve most of the chemical properties of acidic soils: A global meta-analysis. ACS Agricultural Science & Technology. 2024;**4**(3):292-306. DOI: 10.1021/acsagscitech.3c00564

[87] Gautam K, Singh P, Singh RP, Singh A. Impact of rice-husk biochar on soil attributes, microbiome interaction and functional traits of radish plants: A smart candidate for soil engineering. Plant Stress. 2024;**14**:100564

[88] Bagues M, Neji M, Karbout N, Boussora F, Triki T, Guasmi F, et al. Mitigating salinity stress in barley (Hordeum vulgare L.) through biochar and NPK fertilizers: Impacts on physio-biochemical behavior and grain yield. Agronomy. 2024;**14**(2):317. Available from: https://www.mdpi.com/2073-4395/14/2/317/htm

[89] Jaffar MT, Chang W, Zhang J, Mukhtar A, Mushtaq Z, Ahmed M, et al. Sugarcane bagasse biochar boosts maize growth and yield in salt-affected soil by improving soil enzymatic activities. Journal of Environmental Management. 2024;**363**:121418

[90] Zhang S, Xue L, Liu J, Jia P, Feng Y, Xu Y, et al. One–third substitution of nitrogen with cow manure or biochar greatly reduced N_2O emission and carbon footprint in saline–alkali soils. Field Crops Research. 2024;**316**:109517

[91] Wang X, Riaz M, Babar S, Eldesouki Z, Liu B, Xia H, et al. Alterations in the composition and metabolite profiles of the saline-alkali soil microbial community through biochar application. Journal of Environmental Management. 2024;**352**:120033

[92] Wang X, Riaz M, Xia X, Babar S, El-Desouki Z, Li Y, et al. Alleviation of cotton growth suppression caused by salinity through biochar is strongly linked to the microbial metabolic potential in saline-alkali soil. Science of the Total Environment. 2024;**922**:171407

[93] Zhang Y, Miao S, Song Y, Wang X, Jin F. Biochar application reduces saline–alkali stress by improving soil functions and regulating the diversity and abundance of soil bacterial community in highly saline–alkali paddy field. Sustainability (Switzerland). 2024;**16**(3):1001. Available from: https://www.mdpi.com/2071-1050/16/3/1001/htm

[94] Geng H, Wang F, Wu H, Qin Q, Ma S, Chen H, et al. Biochar and nano-hydroxyapatite combined remediation of soil surrounding tailings area: Multi-metal(loid)s fixation and soybean rhizosphere soil microbial improvement. Journal of Hazardous Materials. 2024;**469**:133817

[95] Jia Q, Sun J, Gan Q, Shi N, Fu S. Zea mays cultivation, biochar, and arbuscular mycorrhizal fungal inoculation influenced lead immobilization. Microbiology Spectrum. 2024;**12**(4):e03427-23. DOI: 10.1128/spectrum.03427-23

[96] Chen X, Ran Z, Li R, Duan W, Song Z, Fang L, et al. Biochar reduces the cadmium content of Panax quinquefolium L. by improving rhizosphere microecology. Science of the Total Environment. 2024;**915**:170005

[97] Liu H, Li C, Lin Y, Chen YJ, Zhang ZJ, Wei KH, et al. Biochar and

organic fertilizer drive the bacterial community to improve the productivity and quality of Sophora tonkinensis in cadmium-contaminated soil. Frontiers in Microbiology. 2023;**14**:1334338

[98] Tong X, Song Q, Wang L, Hong Z, Dong Y, Jiang J. Effects of biochars derived from four crop straws on a Cd-polluted cinnamon soil. Environmental Science and Pollution Research. 2023;**30**(9):24764-24770. DOI: 10.1007/s11356-023-25440-8

[99] Dengxiao Z, Hongbin J, Wenjing Z, Qingsong Y, Zhihang M, Haizhong W, et al. Combined biochar and water-retaining agent application increased soil water retention capacity and maize seedling drought resistance in Fluvisols. Science of the Total Environment. 2024;**907**:167885

[100] García Gómez P, Moreno DA, Conesa E, del Martínez Ballesta MC. Effect of biochar amendment and organic fertilization on the yield and nutritional quality of artichoke (Cynara cardunculus L.). Horticulturae. 2024;**10**(9):910. Available from: https://www.mdpi.com/2311-7524/10/9/910/htm

[101] Dong Y, Zhang X, Wang X, Xie C, Liu J, Cheng Y, et al. Modified biochar affects CO_2 and N_2O emissions from coastal saline soil by altering soil pH and elemental stoichiometry. Science of the Total Environment. 2024;**954**:176283

[102] Al Taheri MR, Astaraei AR, Lakzian A, Emami H. The role of biochar and sulfur-modified biochar on soil water content, biochemical properties and millet crop under saline-sodic and calcareous soil. Plant and Soil. 2024;**499**(1-2):221-236. DOI: 10.1007/s11104-023-05912-z

[103] Wang J, Riaz M, Babar S, Xia H, Li Y, Xia X, et al. Iron-modified biochar reduces nitrogen loss and improves nitrogen retention in Luvisols by adsorption and microbial regulation. Science of the Total Environment. 2023;**879**:163196

[104] Xia X, Riaz M, Babar S, Li Y, Wang X, Wang J, et al. Acid-modified cotton straw biochar has instructive for the improvement of saline-alkali soil. Journal of Soils and Sediments. 2024;**24**(6):2334-2348. DOI: 10.1007/s11368-024-03800-2

[105] Gong Y, Hou R, Fu Q, Li T, Wang J, Su Z, et al. Modified biochar reduces the greenhouse gas emission intensity and enhances the net ecosystem economic budget in black soil soybean fields. Soil and Tillage Research. 2024;**237**:105978

[106] Qayyum MF, DESEE K, SMS A, Sakit Alhaithloul HA, Alsudays IM, Rizwan M, et al. Agricultural waste-based modified biochars differentially affected the soil properties, growth, and nutrient accumulation by maize (*Zea mays* L.) plants. BMC Plant Biology. 2024;**24**(1):1-13. DOI: 10.1186/s12870-024-05202-5

[107] Xue B, Huang L, Huang Y, Zhou F, Li F, Kubar KA, et al. Roles of soil organic carbon and iron oxides on aggregate formation and stability in two paddy soils. Soil and Tillage Research. 2019;**187**:161-171

[108] Han L, Sun K, Yang Y, Xia X, Li F, Yang Z, et al. Biochar's stability and effect on the content, composition and turnover of soil organic carbon. Geoderma. 2020;**364**:114184. DOI: 10.1016/j.geoderma.2020.114184

[109] Bi H, Xu J, Li K, Li K, Cao H, Zhao C. Effects of biochar-coated nitrogen fertilizer on the yield and quality of Bok choy and on soil nutrients. Sustainability. 2024;**16**(4):1659.

Available from: https://www.mdpi.com/2071-1050/16/4/1659/htm

[110] Hossain MZ, Bahar MM, Sarkar B, Bolan N, Donne S. Fertilizer value of nutrient-enriched biochar and response of canola crop. Journal of Soil Science and Plant Nutrition. 2024;**24**(2):2123-2137. DOI: 10.1007/s42729-024-01784-z

[111] Kumari R, Kumar V, Kumar A. Soil and biochar: Attributes and actions of biochar for reclamation of soil and mitigation of soil borne plant pathogens [Internet]. Journal of Soil Science and Plant Nutrition. Springer. 2024;**24**(2024):1924-1939. DOI: 10.1007/s42729-024-01810-0

[112] Jin X, Zhang T, Hou Y, Bol R, Zhang X, Zhang M, et al. Review on the effects of biochar amendment on soil microorganisms and enzyme activity. Journal of Soils and Sediments. 2024;**24**(7):2599-2612. DOI: 10.1007/s11368-024-03841-7

[113] Sun C, Liu Y, Bei K, Zheng W, Wang Q, Wang Q. Impact of biochar on the degradation rates of three pesticides by vegetables and its effects on soil bacterial communities under greenhouse conditions. Scientific Reports. 2024;**14**(1):1-14. Available from: https://www.nature.com/articles/s41598-024-70932-3

[114] Rivera-García JA, Díaz T, Contreras-Cornejo HÁ, Larsen J, González-Esquivel CE, López-Pérez L, et al. Biochar alters maize growth, mycorrhizal associations and fall armyworm performance. Plant and Soil. 2023:1-13. DOI: 10.1007/s11104-024-06874-6

[115] Liu S, Li J, Zhou Z, Steinberg CEW, Pan B, Tao S, et al. Biochar soil addition alters ant functional traits as exemplified with three species. Biochar. 2024;**6**(1):1-12. DOI: 10.1007/s42773-024-00337-y

[116] Zhao J, Qiu Y, Yi F, Li J, Wang X, Fu Q, et al. Biochar dose-dependent impacts on soil bacterial and fungal diversity across the globe. Science of the Total Environment. 2024;**930**:172509

[117] Zou G, Zhao F, Lan X, Nawaz M, Shohag JI. Role of coconut Shell biochar on soil properties, microbial diversity and nitrogen mineralization in tropical Latosol. Polish Journal of Environmental Studies. 2024;**33**(2):1487-1496. Available from: https://www.pjoes.com/Role-of-Coconut-Shell-Biochar-on-Soil-Properties-nMicrobial-Diversity-and-Nitrogen,174014,0,2.html

[118] Liu Y, Gao L, Wang C, Fu Z, Chen R, Jiang W, et al. Biochar combined with humic acid improves the soil environment and regulate microbial communities in apple replant soil. Ecotoxicology and Environmental Safety. 2024;**283**:116958

[119] Liu Y, Wang C, Chen R, Jiang W, Li Y, Yin C, et al. Biochar alleviates apple replant disease by reducing the growth of *Fusarium oxysporum* and regulating microbial communities. Horticultural Plant Journal. 2024;**10**(3):657-671

[120] Ahmad S, Zhai X, Wang M, Shi Y, Chen Y, Liang Q, et al. Biochar amendments improve soil functionalities, microbial community and reduce Pokkah boeng disease of sugarcane. Chemical and Biological Technologies in Agriculture. 2024;**11**(1):1-14. DOI: 10.1186/s40538-024-00546-4

Chapter 3

Storage Time in Bokashi Alters the Growth and Fruit Production in Cucumber

Maribel López-Atanacio, Víctor Rosas-Guerrero, Rodrigo Lucas-García, Lorena Alemán-Figueroa and José Antonio Gama-Salgado

Abstract

Even when there are many studies that assess the response of crops to bokashi fertilization, none have evaluated the effect of the storage time of bokashi on the growth, flowering, and yield of crops. In this study, the effect of five different bokashi storage times on vegetative, flowering, and yield traits of cucumber was evaluated. Six treatments were applied, including freshly made bokashi, bokashis with 1, 5, 9, and 13 months of storage, and without any application of bokashi. It was found that the plants fertilized with bokashi with longer storage time increased their production of commercial fruits and yield and decreased the number of non-commercial fruits compared to plants with no application of bokashi and with plants fertilized with freshly made bokashi. These results indicate that the bokashi should be used after a storage period of at least 5 months for greater effectiveness in the production of organic vegetables.

Keywords: *Cucumis sativus*, fruit malformation, natural fertilizers, organic manure, phenological traits, vegetative traits

1. Introduction

The intensive use of synthetic fertilizers has been a common practice in modern agriculture [1]. However, the application of high doses of these inputs can affect people's health and can have negative consequences on ecosystems [2]. It has been documented that their excessive use contributes to groundwater pollution, causing eutrophication and accelerating soil degradation [2, 3]. In addition to this, the dependence on synthetic fertilizers increases production costs, putting at risk the economic sustainability of agricultural systems [4]. Facing these challenges, organic fertilizers have become an alternative within sustainable agriculture [1, 5], since these soil amendments have been shown to improve soil fertility and moisture retention, as well as to contribute to food security, as they offer safer and more nutritious foods [6, 7].

The bokashi is one of these organic fertilizers that is widely used in various agricultural regions and is recognized for its high content of nutrients, beneficial microorganisms, and organic matter [5]. This fertilizer is characterized by activating and increasing beneficial microorganisms in the soil, improving its physical characteristics, and providing nutrients to plants [8]. Unlike other organic fertilizers, such as chicken manure or compost, which require many weeks or months for their decomposition process, respectively [9], the decomposition process of bokashi occurs between 14 and 21 days [5]. In some cases, the final product is a semi-stabilized organic material, that is, a partially decomposed and biochemically unstable material [10]. It has been found that the stabilization of bokashi continues once it is applied to the soil [5]. However, the application of a non-stabilized product could generate a negative impact on plant growth due to a decrease in oxygen supply, immobilization of N in the soil, or phytotoxicity generated by the presence of toxic compounds [5, 11].

A possible solution to mitigate the negative effects of applying an unstabilized organic fertilizer would be to store the product before its use [9, 12], allowing the maturation and decomposition process to continue in a controlled manner [9, 10, 12, 13]. This storage process not only favors the stabilization of the fertilizer but also facilitates the transformation of its organic components into more available forms for plants [9, 10, 12, 14]. It has been found that as storage time increases, organic fertilizers experience greater mineralization of nutrients such as nitrogen, phosphorus, and potassium [9, 10, 12, 15]. For example, Hasan et al. [9] found that composts prepared from cotton stems with 9 months of storage had a higher content of nitrogen, phosphorus, and total potassium than those with zero or 3 months of storage. Hammed et al. [16] found that the total nitrogen, phosphorus, and potassium content was higher in the compost stored for 9 months compared to the compost stored for 0, 3, or 6 months. Furthermore, they reported that the content of heavy metals such as chromium, lead, and nickel was lower in the compost after 9 months of storage [16]. Yang et al. [17] found that the concentration of NO_3^- in sheep manure increased as storage progressed over a 90-day period, though the NH_4^+ content decreased. Cerrato et al. [10] found that the C:N ratio in bokashi stabilized between 45 and 48 days after its preparation, but this depends on the ingredients used for its preparation. This suggests that the decomposition of nutrients in bokashi varies over time [10, 14, 15], indicating that the effects of its application to crops may depend on the storage time. Thus, storing the organic fertilizer for an adequate period could ensure a better assimilation of nutrients by plants, improving their growth and reducing the negative environmental impact that could result from the application of an immature product [11]. According to these results, bokashis with a longer storage time are expected to provide greater benefits for plant growth and yield compared to those with a shorter maturity time [15].

All of the above highlights the importance of identifying the optimal time to apply bokashi to plants to maximize nutrient assimilation and achieve high yields. However, to date, there are no studies in any crop that address the effect of bokashi storage time on their growth and yield. The objective of this study is to evaluate the effect of five bokashis with different storage times on vegetative growth, flowering, and yield (quantity and quality of fruits) of a horticultural crop. In particular, we select the cucumber since it is one of the most important economic and nutritional horticultural crops worldwide, thanks to its high demand in national and international markets [18, 19]. In Mexico, the cucumber is one of the most important vegetables for both export and local consumption [19, 20]. Due to its increasing demand, cucumber exports have increased considerably, which has boosted the use of synthetic fertilizers to improve yields [21–23]. The results of this study are expected to provide valuable information for the optimal use of bokashi as fertilizer to increase the production of vegetables in a more sustainable way.

2. Materials and methods

2.1 Study site

The experiment was conducted from July to October 2017 in an experimental plot (17°12′27.1″N, 100°37′53.3″W, 50 masl) characterized by a warm subhumid climate (Aw), with well-differentiated dry and wet seasons, an average annual rainfall of 1100 mm, and an average annual temperature of 27°C with a maximum of 28°C in the summer and a minimum of 14°C in the winter [24]. Soil was collected in the plot to a depth of 0–30 cm to determine their physical and chemical characteristics before the experiment (**Table 1**). In the previous year, this plot was cultivated with squash (*Cucubita pepo*) and melon (*Cucumis melo*).

2.2 Study species

The cucumber (*Cucumis sativus*) is an annual, creeping, or climbing herbaceous cucurbitaceous plant that varies from 0.6 to 3 m in length [25, 26]. Plants are usually monoecious (i.e., with staminate and pistillate flowers on the same plant), although some modern hybrid cultivars can be andromonoecious [25, 26]. Flowering begins 20–30 days after sowing, and flower anthesis occurs between 5:30 and 7:00 h, with the staminate ones being the first to appear [26]. The ratio of male to female flowers varies from 7:1 to 17:1 [27].

Soil parameter	Value
pH	7.0
Electrical conductivity (dS m^{-1})	0.06
Organic matter (%)	2.2
Total N (%)	0.11
P (ppm)	36
K (meq/100 g^{-1})	0.66
Ca (meq/100 g^{-1})	13
Mg (meq/100 g^{-1})	4
Na (meq/100 g^{-1})	0.05
Texture	clayey
Fe (ppm)	11
Cu (ppm)	1.6
Zn (ppm)	2.25
Mn (ppm)	9

Table 1.
Physical and chemical properties of the soil in the experimental plot.

This crop requires insect pollination for proper fruit development and seed production, with bees being the main pollination agents of this crop [25]. The fruits are oblong, cylindrical, oval, and elongated in shape, fragrant and delicious, rich in nutrients that can be consumed both fresh and processed into pickles [26]. The temperature for optimal development ranges from 25 to 28°C [28]. Nutrient addition and irrigation are important factors that influence fruit growth, yield, and quality [23, 29, 30]. In this study, commercial cucumber seeds of the 'Poinsett' cultivar (Distribuidora Rancho Los Molinos SA de CV, Tepoztlán, Morelos, Mexico) were used.

2.3 Experimental design and crop management

To evaluate the effect of the application of bokashi with different storage times on the growth, flowering, and yield (quantity and quality of fruits) of cucumber, the following six treatments were applied: *B0* – bokashi recently produced (one day of storage); *B1, B5, B9, and B13* – bokashis with 1, 5, 9, and 13 months of storage, respectively; and NB – without application of bokashi, used as a control. The application of bokashi (300 g per plant) was carried out 15 days before sowing (July 1, 2017) [15]. The treatments were arranged in a randomized complete block design, with 30 repetitions, for a total of 180 experimental units. Each experimental unit within a block consisted of one cucumber plant. Each block consisted of an 8 m furrow, with a distance between each experimental unit and between furrows of 1.5 m (**Figure 1**).

The soil for cultivation was removed and cleared of weeds mechanically with the aid of a tractor-drawn disc plow and harrow. Sowing was carried out on July 16, 2017, placing three seeds per hole, and thinning 10 days after sowing, leaving only the largest plant. Each plant was labeled with its respective treatment (**Figure 1**). Irrigation was carried out every third day, using a short furrow irrigation system, a traditional irrigation method widely used by local cucumber producers. Although this irrigation system is considered to provide an uneven distribution of water, recent studies indicate

Figure 1.
Distribution of cucumber plants in the experimental plot.

that short furrows show less variation in dimensions of the furrow and slope compared to longer furrows [31, 32] and suggest that water distribution is uniform in furrows less than 30 m in length (furrow length in our experimental plot was 8 m) [31, 32].

Weeding was manually done by hand-picking of weeds every 5 days throughout the crop cycle. The presence of pests was controlled with the application of chlorpyrifos ethyl 33.80% + permethrin 4.8% every 4 days, while quintozene 30% + thiram 30% was used every 8 days to prevent root rot. The doses used were those recommended by manufacturers.

2.4 Elaboration of bokashi

The following materials were used to make each bokashi: mountain soil (3 sacks), bovine manure (3 sacks), oat straw (3 sacks), ground corn (9 kg), crushed charcoal (9 kg), ash (3 kg), molasses (3 L), yeast (0.25 kg), and water [33]. The molasses and yeast were mixed in a container with water beforehand. The remaining materials were mixed evenly and moistened with the mixture of yeast, molasses, and water. The "fist test" (a simple method to assess soil or compost moisture by squeezing a handful of a sample) was used to verify the ideal humidity. The homogeneous mixture was kept under shade and covered with black polyethylene. The mixture was turned twice daily during the first 7 days by shoveling to prevent the temperature from exceeding 70°C and then once a day (see [34] for a complete explanation). At the end of the 15-day decomposition process, the bokashis were stored in sacks in a concrete warehouse until the day of application. All these materials were collected from the same sites.

2.5 Vegetative traits

Thirty days after sowing, in each of the 30 plants per treatment, the following growth parameters were recorded: (1) *length of the main guide* – measured with a flexometer (Truper, precision: 1 mm) from the base of the plant at ground level to the apex of the guide; (2) *stem diameter* – determined with a digital caliper (Mitutoyo, precision: 0.01 mm), 1 cm above the soil surface; and (3) *leaf area* –estimated with a digital photograph (Samsung Galaxy Table SM-T113NU) of the largest leaf of each plant, using the SigmaScan Pro 5 software.

2.6 Floral traits

The following floral traits were recorded for each plant of each treatment: (1) *days to flowering* – number of days from germination to the appearance of the first staminate and pistillate flower; and (2) *number of flowers* – total number of staminate and pistillate flowers that opened during the entire flowering period.

2.7 Yield traits

Four yield traits were estimated for each plant of each treatment: (1) *number of fruits* – classified as 'commercial' if they were predominantly straight and with a polar diameter greater than 15 cm (**Figure 2A**) and as 'non-commercial' if they were curved fruits with a polar diameter less than 15 cm (**Figure 2B**); (2) *fruit weight* – using a scale (Ohaus, Scout Pro model, precision: 0.1 g) of two fruits per plant; and (3, 4) *polar and equatorial diameter* – measured with a digital caliper (Mitutoyo, precision: 0.01 mm) of two fruits per plant. The two fruits selected

Figure 2.
Characteristics of a commercial fruit (A) and a non-commercial fruit (B) in cucumber plants.

were the first two produced by the plant. All fruits (i.e., commercial and non-commercial) were harvested 10 days after flower anthesis. The commercial yield was determined as the number of commercial fruits per plant multiplied by the average weight of the two selected fruits.

2.8 Statistical analysis

A one-way analysis of variance (ANOVA) was performed to analyze the effects of treatment on main shoot length, stem diameter, leaf area, fruit weight, and polar and equatorial fruit diameter since the data showed a normal distribution and homoscedasticity. To compare days to flowering, a generalized linear model (GLM) with a Poisson distribution was used due to data counts. For the number of staminate and pistillate flowers, the number of commercial fruits, and the number of non-commercial fruits, GLMs with negative binomial distribution were used due to data overdispersion. The estimation of significance of differences of means among treatments was carried out by using Tukey's post-hoc test with a significance level of 0.05. Finally, to analyze the total yield per plant and the commercial yield, a Kruskal–Wallis test was performed since the data did not meet the assumptions of normality and homoscedasticity. All statistical analyses were performed using R version 4.4.1 [35]. For the analysis of variance (ANOVA), the *aov* function of the 'stats' package was used. The GLMs were adjusted with the *glm* function of the 'MASS' package. The Kruskal-Wallis test was run with the *kruskal.test* function of the 'stats' package. The significance of the variables in the models was determined using the *anova* function of the 'car' package. Multiple comparisons between treatments were performed using the 'emmeans' package. In particular, the estimated marginal means were calculated with the *emmeans* function, adjusted by the Tukey method for multiple comparisons (adjust = 'tukey').

3. Results

3.1 Vegetative traits

There were marginal differences among treatments in the length of the main shoot ($F_{5,151}$ = 2.24, P = 0.053). Specifically, the length of the main shoot in B13 was greater than in the NB treatment (**Table 2**). However, for stem diameter ($F_{5,152}$ = 1.28, P = 0.276) and leaf area ($F_{5,151}$ = 1.09, P = 0.366), no significant differences were detected (**Table 2**).

3.2 Floral traits

In all treatments, staminate flowers appeared earlier than pistillate flowers (**Table 3**). The onset of flowering of both staminate (χ^2 = 3.50, P = 0.623) and pistillate flowers (χ^2 = 2.01, P = 0.847) was similar among treatments. The production of staminate (χ^2 = 3.22, P = 0.664) and pistillate flowers (χ^2 = 3.84, P = 0.571) did not show significant differences among treatments (**Table 3**). Plants in all treatments produced approximately 12-fold more staminate than pistillate flowers (**Table 3**).

Treatment	Length of the main guide (cm)	Stem diameter (mm)	Leaf area (cm^2)
NB	165.1 ± 7.77 b	13.7 ± 0.43 a	281.0 ± 18.51 a
B0	187.4 ± 8.37 ab	13.7 ± 0.45 a	334.5 ± 20.28 a
B1	182.6 ± 8.21 ab	14.7 ± 0.45 a	303.3 ± 19.89 a
B5	187.8 ± 7.91 ab	14.5 ± 0.44 a	288.2 ± 19.16 a
B9	176.9 ± 8.73 ab	13.6 ± 0.48 a	308.7 ± 21.14 a
B13	200.9 ± 8.21 a	14.1 ± 0.45 a	331.8 ± 19.89 a

Table 2.
Mean values ± standard error of vegetative traits of cucumber plants grown with bokashi of different storage times. Statistically significant differences among treatments are indicated by different letters.

Treatment	Days to staminate flowering	Days to pistillate flowering	Number of staminate flowers	Number of pistillate flowers
NB	31.8 ± 1.05 a	35.8 ± 1.11 a	291.4 ± 22.40 a	24.7 ± 1.84 a
B0	32.8 ± 1.14 a	36.2 ± 1.20 a	300.6 ± 24.88 a	24.2 ± 1.94 a
B1	32.6 ± 1.12 a	36.7 ± 1.19 a	282.5 ± 22.94 a	23.6 ± 1.86 a
B5	32.7 ± 1.08 a	35.9 ± 1.13 a	280.9 ± 21.98 a	27.5 ± 2.06 a
B9	34.6 ± 1.23 a	37.2 ± 1.27 a	256.0 ± 22.13 a	22.4 ± 1.90 a
B13	32.0 ± 1.11 a	35.0 ± 1.16 a	290.7 ± 23.60 a	24.9 ± 1.95 a

Table 3.
Mean values ± standard error of flowering traits of cucumber plants grown with bokashi of different storage times. Statistically significant differences among treatments are indicated by different letters.

3.3 Yield traits

The total number of fruits per plant was similar among treatments (χ^2 = 3.93, P = 0.558). On the other hand, treatments B5 and B13 produced a higher number

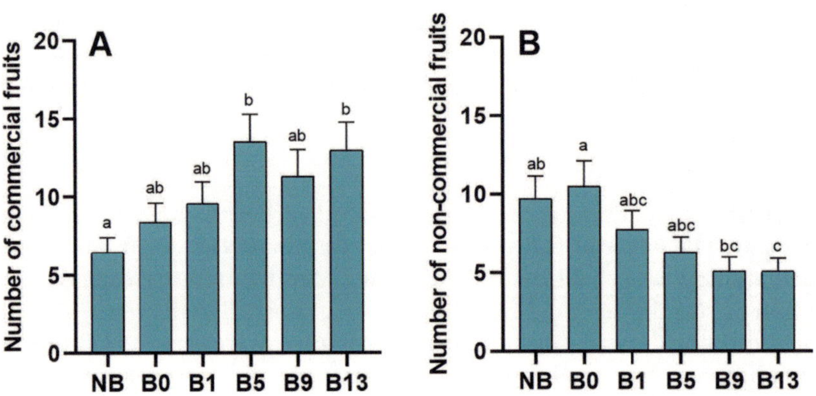

Figure 3.
Number of commercial fruits (A) and non-commercial fruits (B) in cucumber plants grown with bokashi of different storage times. Statistically significant differences among treatments are indicated by different letters.

Treatment	Polar diameter (mm)	Equatorial diameter (mm)	Fresh weight (g)
NB	161.5 ± 3.26 a	51.1 ± 0.82 a	261.6 ± 10.80 a
B0	165.9 ± 3.51 a	50.9 ± 0.88 a	263.9 ± 11.63 a
B1	166.0 ± 3.45 a	50.6 ± 0.87 a	261.3 ± 11.40 a
B5	174.2 ± 3.32 a	52.8 ± 0.84 a	291.5 ± 10.99 a
B9	166.7 ± 3.66 a	50.3 ± 0.92 a	259.9 ± 12.13 a
B13	170.8 ± 3.45 a	51.2 ± 0.87 a	279.7 ± 11.40 a

Table 4.
Mean values ± standard error of fruit traits of cucumber plants grown with bokashi of different storage times. Statistically significant differences among treatments are indicated by different letters.

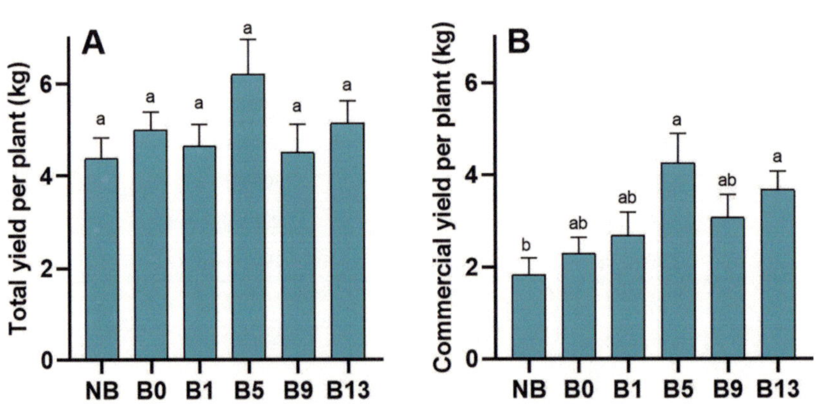

Figure 4.
Total yield (A) and commercial yield (B) of cucumber plants grown with bokashi of different storage times. Statistically significant differences among treatments are indicated by different letters.

of commercial fruits compared to treatment NB (χ^2 = 19.51, P = 0.001; **Figure 3A**). Treatment B0 generated the highest number of non-commercial fruits compared with treatments B9 and B13 (χ^2 = 19.43, P = 0.001; **Figure 3B**).

Regarding fruit quality, no significant differences were observed among treatments in polar diameter ($F_{5,151}$ = 1.76, P = 0.123), equatorial diameter ($F_{5,151}$ = 0.99, P = 0.422), or fresh weight of the fruits ($F_{5,151}$ = 1.34, P = 0.249; **Table 4**).

Total yield was similar between treatments (χ^2 = 4.70, P = 0.452; **Figure 4A**), while plants grown in treatments B5 and B13 had higher commercial yield compared to the NB treatment (χ^2 = 19.49, P = 0.001; **Figure 4B**).

4. Discussion

Our results provide evidence that the application of bokashi with 13 months of storage increased the production of commercial fruits and commercial yield per plant and decreased the number of non-commercial fruits compared to plants with no application of bokashi and with plants fertilized with freshly made bokashi. In contrast, there were no differences among treatments with different storage times in most of the vegetative and floral traits or the quality of commercial fruits.

Previous studies have shown that during storage, organic fertilizers continue their decomposition and stabilization process [9, 10, 12, 13] and increase their nutrient mineralization [9, 10, 12, 15, 36], which facilitates the transformation of their organic components into forms more available to plants [9, 10, 12, 14, 15]. In the present study, it was observed that only plants fertilized with B13 had longer main shoots compared to NB, suggesting that, unlike stem diameter and leaf area, shoot growth requires a greater supply of nutrients. This agrees with Ramírez and Bertsch [37], who found that in cucurbits, shoot production is among the stages with the highest nutritional demand. It is likely that B13 provided the nutrients that the cucumber needed for the growth of the main shoot, especially nitrogen, which is involved in plant growth and in the distribution of assimilated nutrients to vegetative organs [38]. It is also important to consider that cucumber is one of the crops with low nutritional requirements, so it can develop favorably in various types of soil [39]. This could explain why no significant differences were observed in leaf area or stem diameter among treatments. Indeed, Fuentes [40] reported that in the vegetative phase, cucumber uses less than 1% of the whole nutrients, while fruiting represents the stage of greatest nutrient consumption, using around 70%.

Even when it was expected that the application of a fertilizer with a longer storage time would favor flowering, no significant differences were observed among treatments in the time of flowering or in the number of flowers. This result is contrary to that reported by Ruiz [41] who found that cucumber plants where bokashi was applied started flowering earlier than plants without bokashi application. It is possible that the soil used in our study already had enough nutrients to reach flowering at the right time. Alternatively, the onset of flowering and flower production could be more influenced by factors other than nutrients, such as environmental conditions (e.g., temperature, light, and humidity) and hormonal factors that regulate these processes [42].

In our study, bokashi fertilizers with five or more months of storage increased the production of commercial fruits, while the production of non-commercial fruits was reduced. In this context, bokashi with longer storage time could have favored the gradual release of essential nutrients such as nitrogen, phosphorus, and potassium [36], which are necessary for fruit quality [43]. The greater amount of mineralized nitrogen that could have this bokashi may increase the distribution of photoassimilates, favoring a greater production of commercial fruits [29].

The production of commercial fruits may also be related to better pollination [25, 43, 44]. Plants that received bokashi with longer storage time could have been more attractive to pollinators, due to the greater availability of essential nutrients in the plant, which could have improved the quality of floral rewards [43, 45]. For example, Karunakaran et al. [45] found that flowers of almond trees irrigated with high concentrations of P produce greater nectar volume, which caused more honeybee visits, higher fruit set, and higher yield. In this sense, Vaudo et al. [43] found that pistillate cucumber flowers with higher nectar volumes received more visits from bumblebees and had higher fruit set than flowers with lower nectar volumes.

The higher number of non-commercial fruits (i.e., curved fruits with a polar diameter of less than 15 cm) found in treatments without the application of bokashi could be related to the lack of certain nutrients in the soil, such as calcium, which has been considered essential for fruit quality [46]. Chang et al. [47] reported the occurrence of black heart disorder in the cucurbit wax gourd fruit (*Benincasa hispida*) because of calcium deficiency. Moreover, the capacity of plants to absorb calcium increases when the availability of nitrogen increases [48]. Christel [15] reported that NO_3^- concentrations in bokashi increased steadily from 5.83 to 34.31 mg/kg over a four-week storage period. Finally, the presence of phytotoxic compounds, such as heavy metals reported in immature organic fertilizers [5, 11], could have influenced the higher production of non-commercial fruits in B0. These phytotoxins can alter the attractiveness of flowers and nutritional rewards for pollinators [49], causing fewer pollinator visits and a lower number of seeds [50], which may affect the shape and size of cucumber fruits.

Even when there were no differences in fruit quality (i.e., size or weight) among treatments, there were significant differences in the commercial yield per plant, mainly due to the higher number of commercial fruits and to the fewer production of non-commercial fruits of plants by older bokashis. The lower incidence of non-commercial fruits found in plants fertilized with bokashis of longer storage times will have further implications on the economic income of cucumber producers, since larger and non-deformed fruits are more preferred by consumers [51] and have a higher price in the market [52]. Even though this study was carried out for only one season, this study offers new knowledge about a topic that was never tested before. So, future research should cover longer periods to consider the variation in weather conditions and soil fertility due to the long-term transformation of organic materials.

5. Conclusion

The findings of this study suggest that the use of bokashi with longer storage time increases the production of commercial fruits and decreases the production of non-commercial fruits, resulting in a higher commercial yield. Moreover, given that the elaboration of bokashi requires locally available materials, its cost is cheaper compared to commercial fertilizers, which in turn will increase the profitability of this crop. Even when the elaboration of bokashi requires only 2 weeks, it is important to note that it should be used after at least a five-month storage period for greater effectiveness in the production of vegetables such as cucumber.

Acknowledgements

The authors thank Benjamín Ramos-Bravo, Luis David Flores-Valdovinos, Ana Hernández-López, Anayeli Adame-Arroyo, Karen De-la-Cruz, and Guadalupe Ramírez-Ramírez for their help in fieldwork and data collection. José Ángel Vences-Martínez kindly gave us permission to use his land as an experimental plot.

Conflict of interest

The authors declare that they have no conflicts of interest that could influence the publication of this chapter.

Author details

Maribel López-Atanacio[1], Víctor Rosas-Guerrero[1*], Rodrigo Lucas-García[1,2], Lorena Alemán-Figueroa[1] and José Antonio Gama-Salgado[2]

1 Escuela Superior en Desarrollo Sustentable, Universidad Autónoma de Guerrero, Tecpan de Galeana, Mexico

2 Posgrado en Recursos Naturales y Ecología, Facultad de Ecología Marina, Universidad Autónoma de Guerrero, Acapulco de Juárez, Mexico

*Address all correspondence to: victor_rosas@yahoo.com

References

[1] Nobile C, Lebrun M, Védère C, Honvault N, Aubertin ML, Faucon MP, et al. Biochar and compost addition increases soil organic carbon content and substitutes P and K fertilizer in three French cropping systems. Agronomy for Sustainable Development. 2022;**42**(6):119

[2] Jote CA. The impacts of using inorganic chemical fertilizers on the environment and human health. Organic and Medicinal Chemistry International Journal. 2023;**13**(3):555864

[3] Adegbeye MJ, Ravi Kanth Reddy P, Obaisi AI, Elghandour MMMY, Oyebamiji KJ, Salem AZM, et al. Sustainable agriculture options for production, greenhouse gasses and pollution alleviation, and nutrient recycling in emerging and transitional nations - An overview. Journal of Cleaner Production. 2020;**242**:118319

[4] Eisa M, Ragauskaitė D, Adhikari S, Bella F, Baltrusaitis J. Role and responsibility of sustainable chemistry and engineering in providing safe and sufficient nitrogen fertilizer supply at turbulent times. ACS Sustainable Chemistry and Engineering. 2022;**10**(28):8997-9001

[5] Quiroz M, Céspedes C. Bokashi as an amendment and source of nitrogen in sustainable agricultural systems: A review. Journal of Soil Science and Plant Nutrition. 2019;**19**(1):237-248

[6] Assefa S. The principal role of organic fertilizer on soil properties and agricultural productivity -a review. Agricultural Research and Technology: Open Access Journal. 2019;**22**(2):46-50

[7] Popa ME, Mitelut AC, Popa EE, Stan A, Popa VI. Organic foods contribution to nutritional quality and value. Trends in Food Science and Technology. 2019;**84**:15-18

[8] Olle M. Review: Bokashi technology as a promising technology for crop production in Europe. Journal of Horticultural Science and Biotechnology. 2021;**96**(2):145-152

[9] Hasan H, Sardar A, Ashtaputre NM, Mageshwaran V. Storability of bio enriched compost prepared from cotton stalks. Journal of Cotton Research and Development. 2020;**34**(2):301-308

[10] Cerrato ME, Leblanc HA, Cruz W, Genao A. Tiempo de estabilización de bokashis elaborados en fincas de la universidad earth. Tierra Tropical. 2006;**2**(2):161-167

[11] Bernal MP, Alburquerque JA, Moral R. Composting of animal manures and chemical criteria for compost maturity assessment. A Review. Bioresource Technology. 2009;**100**(22):5444-5453

[12] Chastain JP. Impact of storage time on the composition of a finished compost product: A case study. Applied Engineering in Agriculture. 2023;**39**(3):303-313

[13] Pinochet D, Cárdenas C, Carrasco J. ¿Cambia la mineralización del nitrógeno desde abonos orgánicos compost, bokashi y biol provenientes del mismo guano de gallina? Agro Sur. 2017;**45**(2):47-56

[14] Cerrato M, Leblanc H, Kameko C. Potencial de Mineralización de Nitrógeno de Bokashi, Compost y Lombricompost Producidos en la Universidad Earth. Tierra Tropical. 2007;**3**(2):183-197

[15] Christel DM. El uso de Bokashi como una enmienda de fertilidad del suelo en el cultivo de espinaca orgánica [Master Thesis]. Vermont, United States: University of Vermont; 2017

[16] Hammed TB, Laniyan TA, Olanrewaju JA. Nutrient chemical forms, storage effects and Phyto-toxicity of organic fertilizers to selected cereal and legume crops. Compost Science and Utilization. 2024;**31**(3-4):144-153

[17] Yang Y, Chen W, Liu G, Kong Y, Wang G, Yin Z, et al. Effects of cornstalk and sawdust coverings on greenhouse gas emissions during sheep manure storage. Waste Management. 2023;**166**:104-114

[18] Yuan BZ, Bie ZL, Sun J. Bibliometric analysis of cucumber (*Cucumis sativus* L.) research publications from horticulture category based on the web of science. HortScience. 2021;**56**(11):1304-1314

[19] Robledo-Torres V, González-Cortés A, Luna-García LR, Mendoza-Villarreal R, Pérez-Rodríguez MÁ, Camposeco-Montejo N. Histological variations in cucumber grafted plants and their effect on yield. Agronomy. 2024;**14**(7):1377

[20] López-Elías J, Rodríguez JC, Huez MA, Garza SO, Jiménez J, Leyva EI. Producción y calidad de pepino (*Cucumis sativus* L.) bajo condiciones de invernadero usando dos sistemas de poda. Idesia. 2011;**29**(2):21-27

[21] Vasquez-Santiago E, Lira-Saldivar R, Valdéz-Aguilar L, Cárdenas-Flores A, Ibarra-Jiménez L, González-Sandoval D. Respuestas del pepino a la fertilización biológica y mineral con y sin acolchado plástico en condiciones de casa sombra. Revista Internacional de Investigación e Innovación Tecnológica. 2014;**2**(10):1-11

[22] Moreno-Velázquez D, Hernández-Hernández BN, Barrios-Díaz JM, Ibáñez-Martínez A, Cruz-Romero W, Berdeja-Arbeu R. Postharvest quality of cucumber grown with different nutrient solution. Revista Mexicana de Ciencias Agrícolas. 2015;**6**(3):637-643

[23] Bello AS, Huda S, Chen ZH, Khalid MF, Alsafran M, Ahmed T. Evaluation of nitrogen and water management strategies to optimize yield in open field cucumber (*Cucumis sativus* L.) production. Horticulturae. 2023;**9**(12):1336

[24] INEGI (Instituto Nacional de Estadística y Geografía. Prontuario de Información Geográfica Municipal de los Estados Unidos Mexicanos; Clave Geoestadística 12057. Guerrero: Técpan de Galeana; 2009

[25] dos Santos SAB, Roselino AC, Bego LR. Pollination of cucumber *Cucumis sativus* L (Cucurbitales: Cucurbitaceae), by the stingless bees *Scaptotrigona aff. Depili* Moure and *Nannotrigona testaceicornis* Lepeletier (hymenoptera: Meliponini) in greenhouses. Neotropical Entomology. 2008;**37**(5):506-512

[26] Pandey S, Kujur SN. Importance, distribution, botany and genetics. In: Pandey S, Weng Y, Behera TK, Bo K, editors. The Cucumber Genome. 1st ed. Switzerland: Springer International Publishing; 2022. pp. 1-13

[27] Ekeke OC, Agbagwa CA. Breeding biology and effect of pollinators on the fruit characteristics of cucumber (*Cucumis Sativus* L.), Cucurbitaceae. Nigerian Journal of Botany. 2018;**31**(2):325-344

[28] Yang HC, Kim YH, Byun HJ, Choi IL, Vu NT, Kim DH, et al. Identification of appropriate light

intensity and daytime temperature for cucumber (*Cucumis sativus* L.) seedlings in a plant factory with artificial lighting for use as grafting material. Sustainability. 2023;**15**(5):4481

[29] Amer KH, Midan SA, Hatfield JL. Effect of deficit irrigation and fertilization on cucumber. Agronomy Journal. 2009;**101**(6):1556-1564

[30] Parkash V, Singh S, Deb SK, Ritchie GL, Wallace RW. Effect of deficit irrigation on physiology, plant growth, and fruit yield of cucumber cultivars. Plant Stress. 2021;**1**:100004

[31] Karim SM, Karim TH. Enhancing furrow irrigation performance and water productivity through better design and water management in a cracked soil. Iraqi Journal of Agricultural Sciences. 2020;**51**(5):1276-1289

[32] Ashine ET, Bedane MT, Mengesha AA, Admassu Lakewu R, Tadesse KH. Determination of appropriate furrow length and inflow rate for furrow irrigation practice on smallholder farmers at Omo Nada District for maize (*Zea maize* L.) production. Air, Soil and Water Research. 2025;**18**:11786221241308144

[33] Restrepo J. Los abonos orgánicos fermentados. In: El A, B, C de la agricultura orgánica y harina de rocas. 1a ed. Managua, Nicaragua: SIMAS; 2007. p. 262

[34] Secretaría de Agricultura y Desarrollo Rural (SADER). Elaboración Bocachi. In: Manuales prácticos para la elaboración de bioinsumos. Ciudad de México, México: SADER; 2022. p. 25

[35] R Core Team. R: A Language and Environment for Statistical Computing. Vienna, Austria: R Foundation for Statistical Computing; 2024. Available from: https://www.R-project.org/

[36] Ramos-Agüero D, Terry-Alfonso E, Soto-Carreño F, Cabrera-Rodríguez JA. Bocashi: abono orgánico elaborado a partir de residuos de la producción de plátanos en Bocas del Toro, Panamá. Cultivos Tropicales. 2014;**35**(2):90-97

[37] Ramírez F, Bertsch F. Curvas de absorción de nutrimentos en melón (*Cucumis melo*) "Honey Dew" y sandía (*Citrullus lanatus*) "Crimsom Jewel". In: Xl Congreso Nacional Agronómico/ III Congreso Naciolnal de Suelos; 19-23 julio 1999. San José, Costa Rica: MAG; 1999. p. 114

[38] Tafti ARD, Alahdadi I, Nadjafi F, Kianmehr MH, Bagheri M, Akbari GA, et al. Evaluation of different amounts of pelleted manure with urea and micro elements on yield and seed quality of medicinal plant, pumpkin (*Cucurbita pepo* Var. Styriaca). International Journal of Advanced Biological and Biomedical Research. 2014;**2**(10):2663-2677

[39] Andrade-Rosas AC, Yampara-Lovera KF. Comparación de la eficiencia de tres productos orgánicos sobre la productividad del cultivo de pepino (*Cucumis sativus* L.) [Thesis]. Arica, Chile: Universidad de Tarapacá; 2020

[40] Fuentes-Paz ES. Descripción de la dinámica de absorción nutrimental en el cultivo de pepino (*Cucumis sativus* L. hibrido Diomede), bajo condiciones de invernadero en el Centro Experimental Docente de la Facultud de Agronomía (CEDA), Guatemala, C.A. [Thesis]. Ciudad de Guatemala, Guatemala: Universidad de San Carlos de Guatemala; 2015

[41] Ruiz-Espinosa LT. Evaluación de cuatro abonos orgánicos en el cultivo de pepino híbrido thumder (*Cucumis sativus*), en el barrio la Capilla, Parroquia El Tambo, Cantón Catamayo provincia de Loja [Thesis]. Loja, Ecuador: Universidad Nacional de Loja; 2011

[42] Swiader JM, Sipp SK, Brown RE. Pumpkin growth, flowering, and fruiting response to nitrogen and potassium sprinkler fertigation in sandy soil. Journal of the American Society for Horticultural Science. 1994;**119**(3):414-419

[43] Vaudo AD, Erickson E, Patch HM, Grozinger CM, Mu J. Impacts of soil nutrition on floral traits, pollinator attraction, and fitness in cucumbers (*Cucumis sativus* L.). Scientific Reports. 2022;**12**(1):21802

[44] Kiatoko N, Pozo MI, Kasiera W, Kariuki SK, van Langevelde F, Hundt B, et al. Body size as a proxy of probing time and visitation rates on cucumber by two African stingless bees increase fruit quality and seed quantity. Scientia Horticulturae. 2023;**309**:111671

[45] Karunakaran R, Yermiyahu U, Dag A, Sperling O. Phosphorus fertilization induces nectar secretion for honeybee visitation and cross-pollination of almond trees. Journal of Experimental Botany. 2021;**72**(8):3307-3319

[46] Jaime-Guerrero M, Álvarez-Herrera JG, Fischer G. Effect of calcium on fruit quality: A review. Agronomia Colombiana. 2024;**42**(1):1-14

[47] Chang J, Liao D, Li J, Li J, Li Z, Chen X, et al. Calcium deficiency leads to fruit blackheart formation by disrupting glycometabolism and phenylpropanoid metabolism in wax gourd. Postharvest Biology and Technology. 2024;**211**:112851

[48] dos Santos MA, Valadares RV, Neves JCL, da Silva IR, Tótola MR, Costa MD. Ammonium nitrogen increases Ca uptake from non-exchangeable reservoirs by eucalypt plants. Forest Ecology and Management. 2020;**465**:118062

[49] Hoover SER, Ladley JJ, Shchepetkina AA, Tisch M, Gieseg SP, Tylianakis JM. Warming, CO2, and nitrogen deposition interactively affect a plant-pollinator mutualism. Ecology Letters. 2012;**15**(3):227-234

[50] Meindl GA, Ashman TL. Effects of soil metals on pollen germination, fruit production, and seeds per fruit differ between a Ni hyperaccumulator and a congeneric nonaccumulator. Plant and Soil. 2017;**420**(1-2):493-503

[51] Opara UL, Pathare PB. Bruise damage measurement and analysis of fresh horticultural produce-a review. Postharvest Biology and Technology. 2014;**91**:9-24

[52] Hueppe R, Zander K. Perfect apples or sustainable production?—Consumer perspectives from Germany. Journal of Consumer Behaviour. 2024;**23**(2):698-710

Chapter 4

Sustainable Mushroom Cultivation: Exploring Filter Cake Potential as a Biowaste Source for Natural Fertilizer in White Button Mushroom Production

Farzaneh Abdolmaleki, Milad Daneshniya and Mohammad Hossein Maleki

Abstract

Sustainable cultivation can benefit significantly from optimizing biowaste sources like filter cake as natural fertilizers. This study investigates the effects of nine treated substrates (T1–T9) with varying filter cake ratios and salinity percentages on edible mushrooms' performance metrics and physicochemical traits. Key variables driving variability and productivity were identified using Principal Component Analysis (PCA) and correlation analysis. PCA results revealed that base weight and diameter were dominant factors influencing yield, explaining 87.65% of the variance on the F1 axis. Treatments such as T4 showed superior performance in these traits, while specific weight, captured on the F3 axis, distinguished treatments like T7 and T9 for their productivity potential. Chemical composition analysis highlighted a contrast between mineral-dense and moisture-rich treatments; T3 stood out with high levels of dry matter, ash, and sodium, while T2 was notable for its elevated moisture and vitamin content. Correlation analysis further indicated that cap-related traits, including cap weight, pin number, and cap diameter, were the strongest yield determinants, with base traits and specific weight providing additional structural support. This study underscores the value of tailored substrate treatments, with T4, T3, and T9 offering advantages, forming a framework for optimizing productivity and nutritional quality in sustainable mushroom cultivation.

Keywords: filter cake, mushrooms, mineral content, antioxidant activity, biowaste

1. Introduction

Sustainable cultivation practices are essential for addressing global food security challenges while minimizing environmental impacts. Current research spans various crops, including shallots, millets, and cocoa, emphasizing strategies such as organic farming, efficient water management, and integrated pest control [1–3]. These methods aim to enhance productivity, improve yield quality, and maintain ecosystem health. Particularly notable is the role of millets in promoting biodiversity and climate resilience [2]. To develop comprehensive sustainable cultivation plans, researchers employ statistical and mathematical models, considering factors such as arable land selection, cultivation methodologies, and climatic effects [4]. The adoption of sustainable cultivation practices also benefits farmers economically while ensuring long-term food availability and environmental conservation.

Mushroom cultivation, particularly of white button mushrooms (*Agaricus bisporus*), has emerged as a sustainable food source and a means of generating income, especially in developing regions [5, 6]. Mushrooms are nutritionally valuable, being low in salt and sugar yet rich in Vitamin D [5]. However, production challenges remain, notably the high cost and limited nutrient content of commercially available compost used in mushroom cultivation [7]. To address these issues, integrating biowaste and compost in agricultural systems offers promising alternatives.

Biowaste application improves soil health and crop productivity while addressing waste management challenges. Organic amendments derived from biowastes enhance soil properties, increase nutrient availability, and promote plant growth [8, 9]. Additionally, biowastes play a critical role in mitigating climate change by immobilizing heavy metals and sequestering carbon [9]. Processes like anaerobic digestion, composting, and vermicomposting convert industrial and agricultural biowastes—such as coffee grounds, rice straw, and sugarcane bagasse—into valuable soil amendments that improve mushroom substrate properties, leading to higher yields and better quality [10]. Critical factors like water retention, gas exchange, and pH control in cover soil significantly influence mycelial growth and mushroom fruiting success, while high salinity levels can inhibit these outcomes [7, 9]. Optimizing cover soil with organic materials, such as sugarcane filter cake, has shown to enhance soil structure and water retention, making it a sustainable alternative for cultivating mushrooms.

Filter cake, a byproduct of sugar production, holds significant potential as a sustainable fertilizer in agriculture. It can replace inorganic phosphorus and potassium fertilizers, contributing to the conservation of agriculture and resource recycling [7, 11]. Composting or applying filter cake directly enhances phosphorus efficiency and enables the formulation of organomineral fertilizers [11]. Given this potential, the present study was conducted to assess the impact of filter cake added on the ruffling stage on compost performance properties and the resultant mushroom characteristics.

This study added filter cake with compost at varying ratios and salinity levels to produce mushroom substrates. Initially, a pre-test was conducted with 16 treatments to evaluate their impact on mushroom yields, narrowing the number to 8 treatments plus control for the main cultivation phase. After a 50-day cultivation period, the physical properties, performance metrics, and physicochemical properties of the mushrooms were evaluated, highlighting the viability of filter cake as a sustainable casing material for mushroom production.

2. Material and methods

2.1 Preparation of filter cake with different salinity levels

To achieve target salinity levels of 2, 1, and 0.5 dS/m, the filter cake is subjected to a washing process involving repeated cycles of deionized water application and drainage. Spread in 15 cm thick layers on a sloped platform, the filter cake is initially washed with deionized water at a rate of 2 L/kg, with each cycle lasting approximately 15 minutes. For a salinity level of 2 dS/m, three cycles are performed, while five cycles are needed for 1 dS/m, and up to eight cycles are required to reach 0.5 dS/m, with the water application rate reduced to 1–2 L/kg during later stages. After each cycle, the drained water is thoroughly removed to prevent salt reabsorption, and salinity is monitored by measuring the electrical conductivity (EC) of the drainage water in a 1:5 (v/v) ratio through an EC meter (Nanbei, China). Once the washing is complete, the filter cake is air-dried for 48 hours to remove excess moisture, and a final salinity test confirms compliance with the desired threshold.

2.2 Treatments preparation

The treatment scheme forming in this study began with a pre-test to investigate the effects of salinity levels and filter cake ratios on the compost used in mushroom cultivation and the resultant mushroom properties. The study emphasizes the significance of maintaining salinity levels within the optimal threshold of ≤2 dS/m, critical for successful mushroom cultivation. Three salinity levels (0.5, 1, and 2 dS/m) were selected to represent the acceptable range conducive to mushroom growth. Additionally, unwashed filter cake with salinity exceeding 2 dS/m was included to assess the impact of elevated salinity on compost performance and mushroom yield. This approach expands the study's scope to encompass both typical and extreme salinity conditions, providing insights into the resilience of the cultivation system under non-optimal scenarios.

To assess the influence of filter cake added in the ruffling stage of the cultivation, ratios of 10%, 30%, 50%, and 70% filter cake for each salinity level were prepared. Including a low ratio (10%) allows the assessment of minimal filter cake addition, which is less likely to alter the substrate's salinity significantly or overwhelm its structural and nutritional balance. This is a starting point for determining the feasibility of incorporating filter cake. The medium ratios (30% and 50%) provide a mid-range balance, testing scenarios where filter cake constitutes a substantial but not dominant portion of the substrate, which could still retain the beneficial characteristics of the compost while introducing the unique properties of the filter cake. Lastly, the high ratio (70%) represents a scenario where filter cake is the dominant component, pushing the limits of its integration to observe any adverse effects or benefits at extreme concentrations. Including the commercially available compost without any filter cake amendment served as a control (T1), offering a baseline for comparison.

After approximately one month of evaluating mushroom yield across all treatments, the optimal formulations (T2 to T9) were selected to focus on the most promising combinations, ultimately the ones leading to higher mushroom yielding. This strategic refinement ensured the study concentrated on viable options, minimizing resource expenditure on less effective treatments. Overall, this design enables a comprehensive exploration of salinity and filter cake ratios while maintaining practical relevance, aligning to integrate filter cake with commercially available compost for mushroom cultivation (**Table 1**).

2.3 Experimental plan for mushroom cultivation and conduction

The experimental protocol for mushroom cultivation was conducted in a controlled environment with precise regulation of light, relative humidity (85–95%), and

Total experimented treatments	Filter cake added ratio		Salinity	Selected treatment
T1(control)	—	—	—	T1
A1	Unwashed	10%	—	T2
A2	Unwashed	30%	—	T3
A3	Unwashed	50%	—	—
A4	Unwashed	70%	—	—
B1	Washed	10%	2 ds/m	T4
B2	Washed	30%	2 dS/m	T5
B3	Washed	50%	2 dS/m	—
B4	Washed	70%	2 dS/m	—
C1	Washed	10%	1 dS/m	T6
C2	Washed	30%	1 dS/m	T7
C3	Washed	50%	1 dS/m	—
C4	Washed	70%	1 dS/m	—
D1	Washed	10%	0.5 dS/m	T8
D2	Washed	30%	0.5 dS/m	T9
D3	Washed	50%	0.5 dS/m	—
D4	Washed	70%	0.5 dS/m	—

Table 1.
Preparation of experimental treatments, including various percentages of filter cake and salinity levels (treatments T1–T9) added to compost.

temperature (20–25°C). The cultivation area was disinfected using lime sulfur and formalin, and a ventilation system with programmable intervals was installed to regulate carbon dioxide concentrations during the reproductive phase. Experimental treatments, along with the control, were systematically arranged in a grid. The shelving units were thoroughly disinfected before and after placement of the treatments and were enclosed with sealed plastic coverings.

The compost preparation process involved a combination of wheat stubble, chicken manure, urea fertilizer, flour, and gypsum. These components were irrigated, homogenized, and stacked into piles, undergoing a series of turning procedures over two weeks. Urea was introduced on the fourth day to enhance nitrogen content. Compost pasteurization was performed on the 18th day, preparing the compost for inoculation.

Material preparation utilized raw materials, including filter cake sourced from the Qazvin Sugar Company (Qazvin Province, Iran) and compost components (Anzali, Gilan Province, Iran). The filter cake was rinsed, sterilized, and stored in sealed bags until the washing phase. During the planting stage, pre-conditioned and seeded composts were placed in sterilized plastic bags under elevated humidity and arranged on shelving units.

Environmental parameters such as temperature and humidity were meticulously managed throughout the growth phase. A dense white mycelial network developed on the compost surface within two weeks. Subsequently, varying proportions of filter cake were applied, and ruffling was conducted to ensure uniform growth.

The reproductive phase was initiated by reducing the temperature to 17°C, which resulted in the emergence of mushroom primordia (pins) within one week.

The first harvest was completed approximately 50 days after planting. Following each harvest, the ruffled compost-filter cake was replenished with water, and relative humidity was maintained at 95–98%. Two additional flushes were obtained at intervals of five days and one week, respectively, ensuring sustained yields over the cultivation cycle.

2.4 Analysis of physicochemical and performance metrics of mushrooms

2.4.1 Chemical analyses

A set of standard approaches with required modifications was implied to evaluate the trace elements and macronutrient analysis [12–16]. The trace elements, including manganese (Mn), copper (Cu), zinc (Zn), and iron (Fe), were quantified using atomic absorption spectrophotometry (AAS) (Varian Techtron, Model 1200). Samples were digested in concentrated HNO_3 and H_2O_2 using a microwave digestion system. Measurements were taken at 279.5 nm for Mn, 324.8 nm for Cu, 213.9 nm for Zn, and 248.3 nm for Fe, with the AAS instrument calibrated using certified standard solutions. To enhance detection limits, solutions were concentrated twentyfold by hot plate evaporation. Macronutrient analysis included nitrogen (N), phosphorus (P), and potassium (K). Phosphorus was quantified using the molybdenum blue method, with samples analyzed at 880 nm using a UV-visible spectrophotometer. Potassium levels were measured using a flame photometer after preparing a 1:9 (w/v) sample-to-water mixture, filtering, and analyzing a 100 mL aliquot. All instruments were calibrated with certified standards, and stringent quality control protocols, including blanks, duplicates, and reference materials, ensured the reliability of the results.

2.4.2 Performance metrics measurement

The number of mushrooms harvested per replicate was determined using average calculation methods to ensure a consistent and representative measurement of production. This step provided the foundational data required for further yield analysis and other performance metrics. Accurate enumeration of mushrooms is critical for comparing the productivity of different treatment groups. The diameter of the base and cap were measured using an engineering ruler, providing dimensional data for morphological characterization. Additionally, the weights of the mushroom base and cap were individually measured using a digital scale. These measurements offer insights into the growth characteristics and potential quality of the mushrooms for culinary and industrial applications.

2.4.3 Yielding assessment

The performance, or yield, of the mushrooms was calculated as a percentage based on the fresh weight of the mushrooms relative to the fresh weight of the substrate, according to [17] with modification. The formula is as follows:

$$\text{Yield}\ (\%) = \frac{W_F}{W_S} \times 100 \tag{1}$$

Where W_F is the fresh weight of mushrooms, and W_S is the fresh weight of the substrate. This calculation is essential for assessing substrate utilization efficiency in mushroom production and enables comparisons between different treatment conditions.

2.4.4 Dry matter measurement

The percentage of dry matter in the mushrooms was measured by drying the samples at 75°C for 48 hours in a drying oven and then weighing them using a digital scale [18]. The formula is as follows:

$$\text{Dry Matter (\%)} = \frac{\mathrm{DW_S}}{\mathrm{F_W}} \times 100 \quad (2)$$

Where DW_S is the dry weight of the sample, and F_W is the fresh weight of the sample. Dry matter analysis is critical for evaluating the moisture content and determining the stability and storability of the mushrooms.

2.4.5 Biological efficiency evaluation

Biological efficiency was calculated as the ratio of the fresh weight of the mushrooms to the dry weight of the substrate, expressed as a percentage according to [19] with modifications. The formula is as follows:

$$\text{Biological Efficiency (\%)} = \frac{W_F}{DW_S} \times 100 \quad (3)$$

Where W_f is the fresh weight of the whole product, and DW_s is the dry weight of the substrate. This parameter indicates how effectively the substrate was converted into mushroom biomass, serving as a key indicator of cultivation success.

2.4.6 Antioxidant activity measurement

The antioxidant capacity of the mushrooms was assessed using the DPPH free radical scavenging assay [20]. Mushroom extracts were prepared using ethanol to isolate bioactive compounds. The mushroom extract was mixed with the DPPH solution, initiating a reaction where antioxidants in the extract reduced the DPPH radicals, causing a measurable decrease in absorbance. The mixture was incubated in the dark at room temperature for 30 minutes to allow the reaction to proceed wholly. After incubation, the absorbance of the mixture was recorded at 515 nm using a spectrophotometer. Antioxidant activity was calculated using the formula:

$$\text{Antioxidant activity (\%)} = \frac{A_S - A_C}{A_C} \times 100 \quad (4)$$

Where A_S is sample absorption, and A_C is control absorption. This assay provides a quantitative measure of the health-promoting potential of the mushrooms.

2.4.7 Ash measurement

The ash percentage was determined by incinerating the mushroom samples in an electric furnace at 550°C for 8 h according to [21] with modifications. The formula used is as follows:

$$\text{Ash Percentage } (\%) = \frac{W_A}{W_D} \times 100 \tag{5}$$

Where W_A is the weight of ash, and W_D is the weight of dry matter. This analysis is important for evaluating the mineral residue and inorganic content of the mushrooms.

2.4.8 Protein measurement

The total nitrogen content was determined using the macro Kjeldahl method [22]. A mixture of 5 g sample, 10 g potassium sulfate, 0.5 g mercury oxide, and 20 ml sulfuric acid was heated until transparent, followed by 1.5 hours of boiling. After cooling, the mixture was diluted with distilled water and combined with boric acid and indicator drops. The solution was treated with caustic soda and distilled, then titrated with 0.1 N hydrochloric acid. Nitrogen content was calculated, and protein was determined by multiplying nitrogen by 6.38.

$$\text{Nitrogen} = \frac{(V_{as} - V_{ac}) \times A_N \times 1.40}{W_S} \tag{6}$$

$$\text{Nitrogen} = \text{Total nitrogen } (\%) \times 6.25 \tag{7}$$

Where V_{as} is the volume of acid utilized for the sample, V_{ac} is the volume of acid utilized for the control, A_N is the normality of acid, and W_S is the sample weight.

2.4.9 Moisture measurement

Visible and Near-Infrared (Vis–NIR) Spectroscopy was used to measure moisture in mushrooms by analyzing the light reflected from the sample's surface in the visible (400–700 nm) and near-infrared (700–2500 nm) regions. Water's strong absorption at specific NIR wavelengths provides a spectral fingerprint that correlates with moisture content using the Partial Least Squares Regression (PLSR) calibration model. This technique avoids sample damage, requires no chemicals, and delivers results in seconds, making it ideal for quality control in mushroom harvesting, storage, and processing, though it necessitates initial calibration and is sensitive to surface conditions [23].

2.5 Statistical analysis

Statistical analysis was conducted on a dataset of 9 observations, each representing the mean of three replicates. Principal Component Analysis (PCA) was performed using XLSTAT (ver. 2022), with standardization carried out using the $n - 1$ approach to ensure equal contribution of all variables regardless of their original units. The PCA model was constructed based on the Pearson correlation matrix, with a maximum of five factors considered to explore the dimensional structure of the dataset. Significant components were identified based on eigenvalues greater than 1 (Kaiser's criterion), and variables with loadings exceeding ±0.5 were considered to contribute

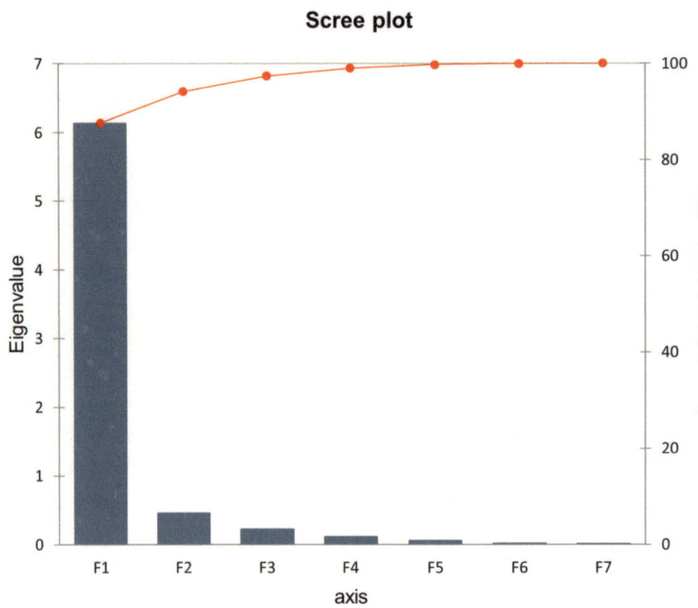

Figure 1.
The relationship between eigenvalues, percentage of explained variance, and cumulative explained variance of the principal component analysis (PCA) model for the mushrooms' physical properties and performance metrics.

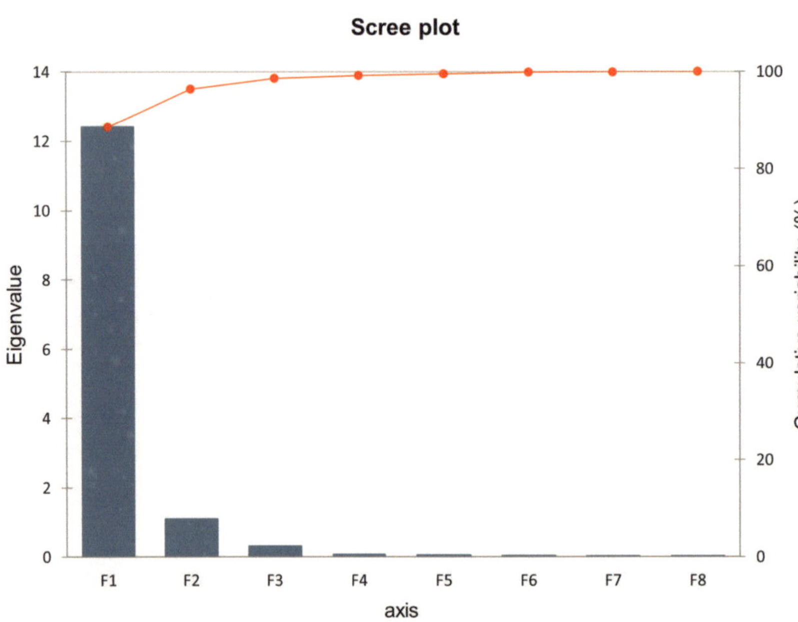

Figure 2.
The relationship between eigenvalues, percentage of explained variance, and cumulative explained variance of the principal component analysis (PCA) model for the mushrooms' physicochemical properties.

meaningfully to the components. A significance level of α = 0.05 was used for interpreting the results, ensuring statistical validity. Below, the relation between the eigenvalues and cumulative variance (%) for the PCA model of morphological and yield characteristics of the mushrooms, as well as their physicochemical properties, can be seen. Regarding these results, only F1 against F2 and F2 against F3 were considered to include solely high variance (**Figures 1** and **2**).

3. Results and discussion

3.1 Impact of different substrate treatments on physical properties and performance metrics of edible mushrooms

The provided PCA biplots represent the relationships between variables, including pins number (per m^2), base and cap diameter (mm), base and cap weight (mm), a specific weight (g) and yielding (%) and treatments (T1–T9). These visualizations help to understand the variability in the dataset and the association between treatments and active variables. The biplots analyze two principal component combinations, F1 and F2 (94.21% of total variance).

In **Figure 3**, which explains 94.21% of the variance, the F1 axis accounts for a dominant 87.65% of the total variance and is heavily influenced by base weight and diameter, as indicated by their long and aligned vectors in the positive F1 direction. These variables are strongly correlated, suggesting that treatments with higher F1 scores are associated with more enormous base weight and diameter values. Other variables, such as cap weight, specific weight, cap diameter, and yielding, also contribute to F1, but their shorter vectors reflect a weaker influence compared to the primary contributors. The F2 axis, explaining 6.55% of the variance, has minor influences from variables

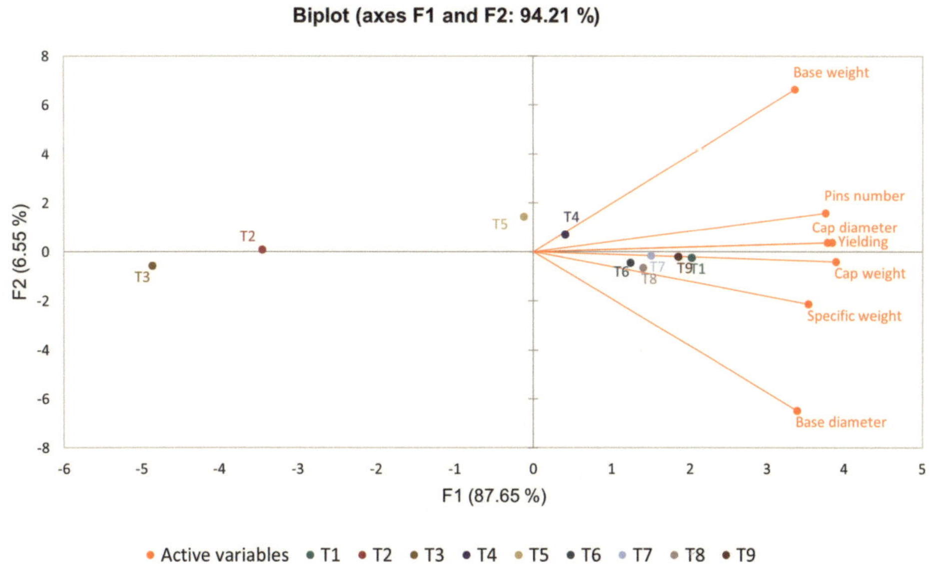

Figure 3.
Biplot of the first two principal components (F1 and F2), accounting for 94.21% of the total variance, highlighting the relationships among variables—including pin number (per m^2), base and cap diameter (mm), base and cap weight (g), specific weight (g), and yielding (%)—in relation to the applied treatments (T1–T9).

like pin number, yielding, and cap diameter, which show some alignment with this axis. However, their contributions remain modest relative to the dominance of F1.

Treatment-wise, T4 is distinctly positioned in the positive F1 and F2 quadrants, strongly associated with base weight and, to a lesser extent, with F2-related variables. This indicates that T4 exhibits the highest values for these characteristics. On the opposite end, T3 and T5 are located far to the negative side of F1, reflecting a negative association with base weight and diameter and suggesting smaller values for these traits compared to other treatments. T6, T7, T8, and T9 are clustered near the origin, signifying average or neutral values across most variables, with no strong association with any specific variable. Meanwhile, T2, positioned slightly to the left of the origin on F1, shows a mild negative association with base weight and diameter, though it is not distinctly different from other treatments.

The interpretation of the axes reveals that F1 is the primary dimension for separating treatments based on base weight and diameter, with treatments scoring high on the right and low on the left. F2, while less impactful, introduces a secondary layer of variability modestly influenced by variables like pin number and yielding. Together, these axes clearly distinguish treatments based on their key characteristics, particularly those linked to weight and size.

In **Figure 4**, which focuses on the F2 and F3 axes, a smaller portion of the variance is explained (6.55% by F2 and 3.19% by F3), yet it reveals unique patterns and subtler behaviors among the treatments. Specific weight emerges as the dominant contributor to F3, as evidenced by its long vector in the positive F3 direction, suggesting that treatments with high F3 scores are strongly associated with high specific weight values. The base weight vector aligns with F2, reinforcing its importance in explaining variability along this axis, as seen in the first biplot, though it contributes moderately to F3. Variables such as cap weight, cap diameter, and yielding are clustered near the origin,

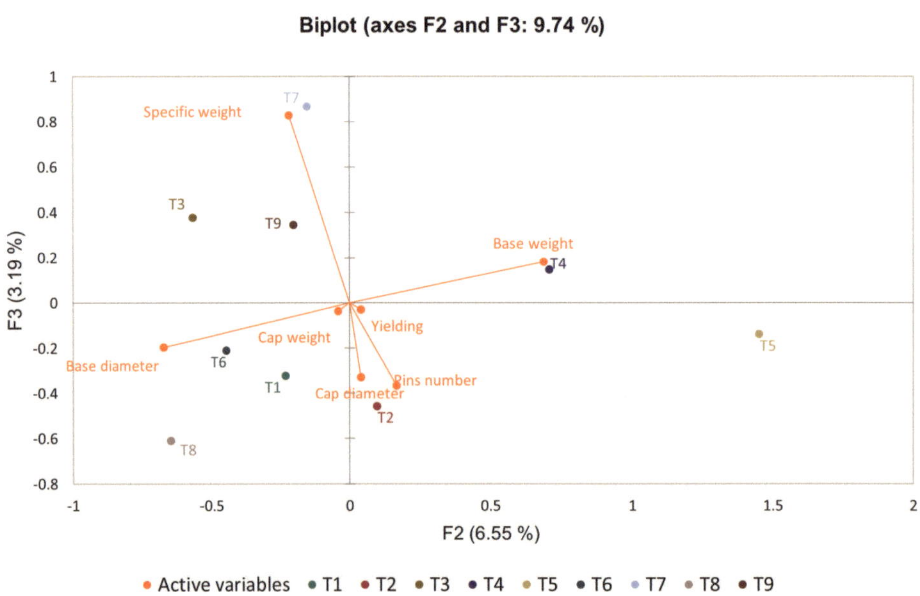

Figure 4.
Biplot of the second and third principal components (F2 and F3), accounting for 9.74% of the total variance, highlighting the relationships among variables—including pin number (per m^2), base and cap diameter (mm), base and cap weight (g), specific weight (g), and yielding (%)—in relation to the applied treatments (T1–T9).

indicating minimal influence on treatment differentiation along either axis. Meanwhile, base diameter shows a slight negative contribution to F3, reflected in its short vector.

Regarding treatment positions, T7 and T9 stand out with high positive F3 scores, closely associated with high specific weight values, which sets them apart from the other treatments. T5 occupies the positive F2 but low F3 region, suggesting alignment with base weight while exhibiting lower specific weight values. In contrast, treatments T6, T1, and T8 are grouped in the negative F3 region, indicating similar profiles characterized by lower Specific weight and neutrality along F2. T3, with a moderate positive F2 score, reflects a weak association with variables contributing to this axis but shows minor F3 contributions.

The interpretation of axes highlights that F2 retains its role in differentiating treatments based on Base weight and related characteristics, consistent with Biplot 1. However, F3 introduces a new dimension, capturing variability driven predominantly by Specific weight, effectively separating treatments like T7 and T9, which exhibit high values for this variable, from others with lower or average profiles. This additional perspective complements the insights gained from the first biplot by emphasizing differences in specific weight.

The two biplots reveal key drivers of variability and distinct treatment behaviors. F1, explaining most variance, is dominated by base weight and diameter, separating treatments like T4 (high values) from T3 and T5 (low values). F2 adds differentiation through variables like pin number and yielding, while F3 highlights Specific weight as a unique factor, distinguishing treatments like T7 and T9.

Treatments near the origin, such as T6, T7, T8, and T9, exhibit balanced traits, though T7 and T9 stand out on F3 due to their high specific weight. Distinct treatments like T4, T3, and T5 are defined by their firm or weak associations with base weight and diameter.

These findings emphasize the primary importance of base weight and diameter in treatment differentiation, with specific weight playing a secondary but significant role. The biplots provide a clear overview of the relationships, highlighting practical implications for optimizing treatments based on their trait profiles.

In addition, **Table 2** shows the Pearson correlation (n-1) table of the variables. The correlation matrix highlights the relationships between various variables influencing

Variables	Pin number	Base diameter	Cap diameter	Base weight	Cap weight	Specific weight	Yielding
Pins number	**1**	0.775	0.950	0.835	0.962	0.796	0.967
Base diameter	0.775	**1**	0.838	0.562*	0.876	0.803	0.826
Cap diameter	0.950	0.838	**1**	0.831	0.959	0.819	0.938
Base weight	0.835	0.562*	0.831	**1**	0.838	0.727	0.837
Cap weight	0.962	0.876	0.959	0.838	**1**	0.897	0.991
Specific weight	0.796	0.803	0.819	0.727	0.897	**1**	0.882
Yielding	0.967	0.826	0.938	0.837	0.991	0.882	**1**

**Values are not significantly different from zero at a significance level of α = 0.05*

Table 2.
The Pearson correlation (n-1) table revealing the relationships among pin number, base diameter, cap diameter, base weight, cap weight, specific weight, and yielding.

yield and crop characteristics. Yield is most strongly influenced by cap weight (0.991), pin number (0.967), and cap diameter (0.938), indicating that maximizing cap size, weight, and pin formation significantly enhances productivity.

Cap-related traits are highly interdependent, with cap weight strongly correlated with cap diameter (0.959) and pin number (0.962), underscoring their collective role in yield improvement. Additionally, base characteristics like base diameter (0.826) and base weight (0.837) moderately contribute to yield, reflecting their supportive role in accommodating pins and increasing cap size. Specific weight (0.882) is another key factor, showing that denser specimens are associated with higher yields and better overall crop performance. The matrix also reveals the interdependence between pin number and cap traits, as more pins correlate with larger caps (0.950) and heavier caps (0.962). Although base traits are less influential than cap characteristics, their positive correlations with yield indicate their importance in structural support. These findings emphasize the central role of cap-related characteristics and pin numbers in driving yield, with base traits and specific weight playing supporting roles.

3.2 Impact of different substrate treatments on physicochemical properties of edible mushrooms

The provided PCA biplots represent the relationships between variables, including moisture, dry matter, biological Efficiency, ash, nitrogen, protein and antioxidant capacity (all measured in %), as well as iron, zinc, sodium, potassium, phosphor, B2 (riboflavin) B3 (niacin) (all measured in mg/50 g), and the treatments (T1–T9). These visualizations help to understand the variability in the dataset and the association between treatments and active variables. The biplots analyze two principal component combinations, F1 and F2 (96.44%of total variance) in the first plot and F2 and F3 (10.03% of total variance) in the second.

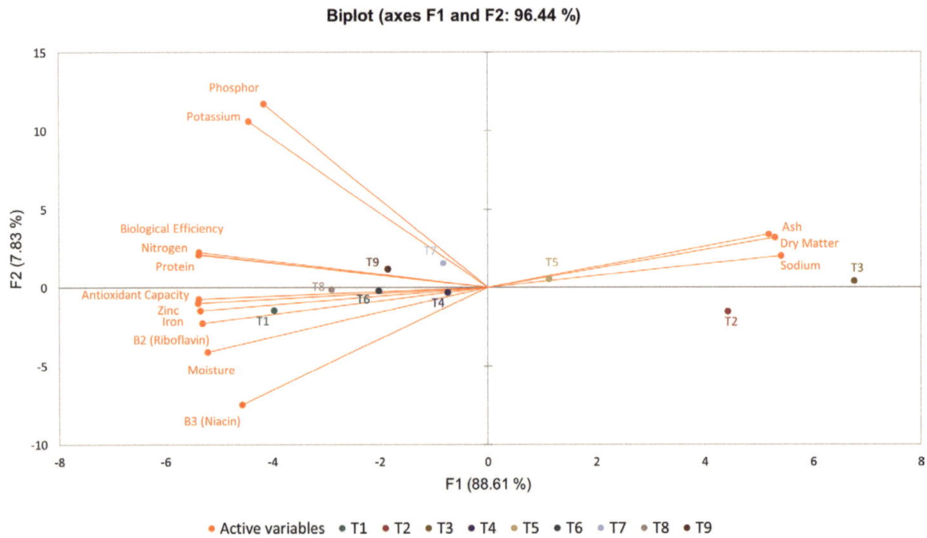

Figure 5.
Biplot of the first two principal components (F1 and F2), accounting for 96.44% of the total variance, highlighting the relationships among variables—including moisture, dry matter, biological efficiency, ash, nitrogen, protein and antioxidant capacity all measured as %, as well as iron, zinc, sodium, potassium, phosphorus, vitamin B2 (riboflavin), vitamin B3 (niacin) all measured in mg/50 g—in relation to the applied treatments (T1–T9).

In **Figure 5**, the active variables are distributed across positive and negative directions along F1 and F2. Variables such as dry matter, ash, and sodium are strongly positively correlated with F1, indicating that these variables increase together. Similarly, phosphorus and potassium are positively aligned with F1 and have some contribution to F2, suggesting a relationship with other nutrient content. In contrast, variables like moisture and B3 are strongly negatively correlated with F1, signifying an inverse relationship between dry matter and sodium. Other variables such as antioxidant capacity, iron, zinc, and B2 are moderately negatively aligned with F1, showing an opposite trend compared to mineral-heavy variables.

The vertical axis (F2) is dominated by phosphorus and potassium in the positive space, while biological efficiency, nitrogen, and protein cluster closer to the origin. Moisture and vitamins such as B3 are mostly neutral along F2, showing stronger influences on F1. These variable groupings provide insights into how treatments perform regarding nutrient distribution and productivity.

When examining the treatments, T3 is positioned farthest along positive F1, suggesting it has the highest dry matter, ash, and sodium levels, making it a mineral-rich treatment. On the other hand, T2 is located firmly on the negative side of F1, indicating it is high in Moisture and B3, setting it apart from the mineral-rich treatments. T9 aligns closely with biological efficiency, nitrogen, and protein, suggesting it is highly productive and rich in protein. T7 is distinct, located in a quadrant without any strong associations, while T1, T4, T6, and T8 cluster near the origin, reflecting balanced profiles without strong dominance in any variable. Finally, T5 is positioned near ash, dry matter, and sodium but not as prominently as T3, indicating moderate mineral content.

Figure 6 focuses on F2 and F3, which capture finer variations in the dataset. Along F2, variables such as biological efficiency, nitrogen, and protein are positively correlated, mirroring their positions in the first biplot. Dry Matter, ash, and sodium also

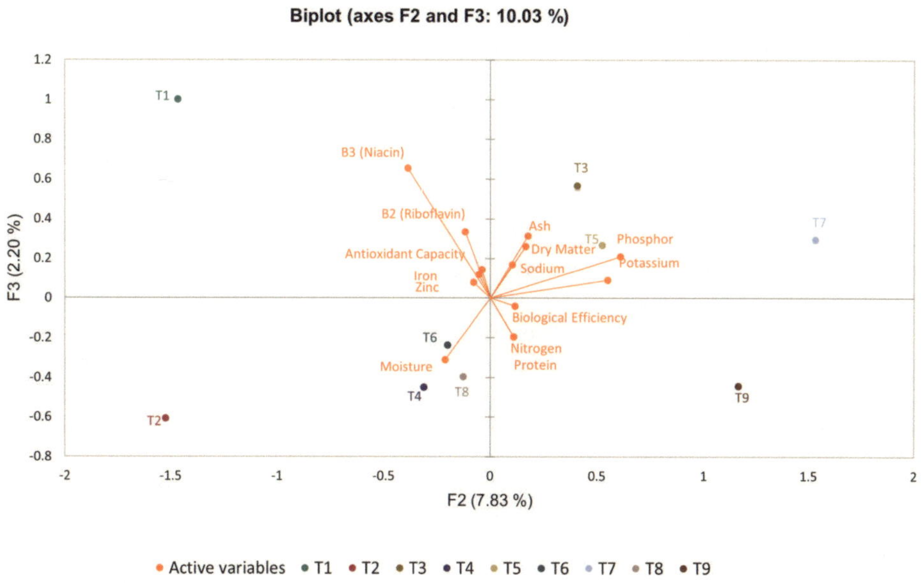

Figure 6.
Biplot of the first two principal components (F2 and F3), accounting for 10.03% of the total variance, highlighting the relationships among variables—including moisture, dry matter, biological efficiency, ash, nitrogen, protein and antioxidant capacity all measured as %, as well as iron, zinc, sodium, potassium, phosphorus, vitamin B2 (riboflavin), vitamin B3 (niacin) all measured in mg/50 g—in relation to the applied treatments (T1–T9).

remain aligned with the positive side of F2, reinforcing their mineral dominance. On the negative side of F2, variables like moisture, antioxidant capacity, and vitamins such as B2 and B3 appear, further supporting their contrasting relationships with the mineral-heavy variables.

The vertical axis (F3) provides additional insights into the behavior of vitamins. Both B3 and B2 contribute positively to F3, indicating their association with treatments rich in vitamins. Moisture is negatively aligned with F3, distinguishing treatments with high moisture from those with high vitamin content. This separation highlights subtle differences between treatments that were less evident in the first biplot.

In terms of treatments, T1 is positioned high along F3, showing strong alignment with vitamins like B2 and B3. T2 remains negatively aligned with both F2 and F3, confirming its distinctiveness as a moisture-heavy treatment with lower mineral and vitamin content. T3 continues to align with ash, dry matter, and sodium, cementing its profile as a mineral-dominant treatment. T9, on the other hand, remains firmly aligned with biological efficiency, protein, and nitrogen, highlighting its productivity-oriented traits. Treatments such as T4, T6, and T8 cluster near the origin, showing balanced profiles, while T7 is positioned distinctly along positive F3, highlighting unique characteristics that are not strongly tied to specific variables.

A comparison of the two biplots highlights several consistent patterns and key distinctions. Enormously influential variables such as phosphorus and potassium remain prominent across both plots, indicating their critical role in defining variation in the dataset. Similarly, moisture, B3, and B2 consistently oppose mineral-heavy variables like ash and sodium, reflecting a clear dichotomy in the dataset. These relationships indicate that treatments can be categorized based on moisture/vitamin richness versus mineral dominance.

Distinct treatments such as T3 and T2 show consistent alignments across both biplots. T3 stands out as mineral-rich, with high levels of dry matter, ash, and sodium, while T2 is moisture-rich, with higher vitamin content. T9 consistently emerges as a productivity-oriented treatment, aligning with variables like biological efficiency and Protein. T1, on the other hand, shows prominence in vitamins, as emphasized in the F2-F3 biplot.

Balanced treatments such as T4, T6, and T8 cluster near the origin in both plots, indicating moderate, neutral profiles. T7 is consistently positioned uniquely, suggesting the presence of unexplored or niche traits. These patterns confirm that PCA effectively captures the relationships among variables and treatments, providing a robust framework for analyzing agricultural or nutritional optimization.

In addition, **Table 3** shows the Pearson correlation (n-1) table of the variables. The correlation matrix highlights the relationships between various factors influencing chemical composition, biological efficiency, and antioxidant capacity. Antioxidant capacity is most strongly influenced by nitrogen (0.951), protein (0.951), and riboflavin (B2) (0.971), indicating that optimizing nitrogen content, protein levels, and vitamin B2 can significantly enhance antioxidant properties.

Protein-related traits are highly interdependent, with nitrogen and protein perfectly correlated (1.000), emphasizing their collective role in improving biological efficiency and antioxidant capacity. Moisture shows a strong positive correlation with antioxidant capacity (0.932) and riboflavin (0.924), but a negative correlation with dry matter (−0.996) and sodium (−0.984), highlighting the trade-offs between moisture levels and these components. Additionally, ash content exhibits a strong correlation with dry matter (0.960) and biological efficiency (−0.929), reflecting its dual influence on nutritional composition and productivity.

Variables	Moisture	Dry matter	Biological efficiency	Ash	N	Protein	Fe	Zn	Na	K	P	B2	B3	Antioxidant Capacity
Moisture	**1**	−0.996	0.905	−0.966	0.935	0.935	0.934	0.942	−0.984	0.641*	0.568*	0.924	0.830	0.932
Dry matter	−0.996	**1**	−0.921	0.960	−0.951	−0.951	−0.947	−0.956	0.986	−0.681	−0.612*	−0.933	−0.832	−0.939
Biological efficiency	0.905	−0.921	**1**	−0.929	0.982	0.982	0.960	0.952	−0.956	0.869	0.816	0.939	0.761	0.962
Ash	−0.966	0.960	−0.929	**1**	−0.917	−0.917	−0.925	−0.922	0.963	−0.663*	−0.578*	−0.915	−0.802	−0.935
N	0.935	−0.951	0.982	−0.917	**1**	1.000	0.956	0.959	−0.972	0.858	0.809	0.922	0.737	0.951
Protein	0.935	−0.951	0.982	−0.917	1.000	**1**	0.956	0.959	−0.972	0.858	0.809	0.922	0.737	0.951
Fe	0.934	−0.947	0.960	−0.925	0.956	0.956	**1**	0.978	−0.972	0.740	0.694	0.971	0.857	0.960
Zn	0.942	−0.956	0.952	−0.922	0.959	0.959	0.978	**1**	−0.979	0.766	0.725	0.971	0.863	0.988
Na	−0.984	0.986	−0.956	0.963	−0.972	−0.972	−0.972	−0.979	**1**	−0.732	−0.670	−0.954	−0.833	−0.974
K	0.641*	−0.681	0.869	−0.663*	0.858	0.858	0.740	0.766	−0.732	**1**	0.990	0.728	0.467*	0.781
P	0.568*	−0.612*	0.816	−0.578*	0.809	0.809	0.694	0.725	−0.670	0.990	**1**	0.681	0.418*	0.731
B2	0.924	−0.933	0.939	−0.915	0.922	0.922	0.971	0.971	−0.954	0.728	0.681	**1**	0.930	0.971
B3	0.830	−0.832	0.761	−0.802	0.737	0.737	0.857	0.863	−0.833	0.467*	0.418*	0.930	**1**	0.863
Antioxidant Capacity	0.932	−0.939	0.962	−0.935	0.951	0.951	0.960	0.988	−0.974	0.781	0.731	0.971	0.863	**1**

Values are indifferent from 0 with a significance level of a = 0.05.

Table 3.
The Pearson correlation (n-1) matrix showing the relationships among pin number, base diameter, cap diameter, base weight, cap weight, specific weight, and yield.

Phosphorus and potassium display moderate positive correlations with biological efficiency (0.869 and 0.866, respectively), emphasizing their supportive roles in overall crop performance. While sodium negatively correlates with biological efficiency (−0.972) and antioxidant capacity (−0.974), this highlights potential inhibitory effects. These findings underscore the central role of nitrogen, protein, and riboflavin in enhancing antioxidant properties, with moisture, ash, and mineral content playing important complementary roles.

4. Conclusion and prospects

This study comprehensively analyses the impact of different substrate treatments (T1–T9) on the physical, chemical, and productivity characteristics of edible mushrooms using principal component analysis (PCA) and Pearson correlation metrics. The findings underscore the pivotal role of base weight and diameter in driving variability among treatments, as indicated by their dominant contributions to the F1 axis, explaining a substantial proportion of the total variance. Treatments such as T4 emerge as significant due to their association with enhanced base weight and diameter, suggesting their potential for optimizing yield traits. Similarly, specific weight, highlighted on the F3 axis, provides an additional layer of differentiation, with treatments like T7 and T9 exhibiting superior values in this regard.

Chemical and physicochemical analyses reveal apparent dichotomies between treatments emphasizing mineral-rich profiles, as seen in T3, and those favoring moisture and vitamin content, exemplified by T2. Variables such as dry matter, ash, and sodium display strong positive associations, contrasting with the trends observed for moisture, B2, and B3, emphasizing trade-offs between mineralization and hydration/vitamin concentration. Furthermore, treatments like T9, aligning with high biological efficiency, protein, and nitrogen, demonstrate promising profiles for productivity optimization.

Correlation analysis corroborates these findings, highlighting the central roles of nitrogen, protein, and antioxidant capacity as key factors driving productivity and nutritional quality. The interdependence between cap-related traits, such as weight and diameter, with yield performance underscores their collective role in achieving higher crop outputs. Simultaneously, the complementary yet opposing influences of moisture and mineral content suggest the necessity of a balanced approach in substrate treatment to maximize overall mushroom quality.

In conclusion, the results demonstrate that substrate treatments significantly affect mushroom growth, nutrient composition, and productivity. Distinct treatments, including T4, T3, and T9, offer unique advantages depending on the desired outcome, whether focused on physical robustness, nutrient enrichment, or productivity. The dual analytical approach of PCA and Pearson correlation provides a robust framework for evaluating these relationships, paving the way for informed strategies to optimize substrate formulations. This study highlights the potential for tailored treatments to enhance agricultural outputs and meet specific nutritional and productivity goals in mushroom cultivation systems.

Conflict of interest

The authors declare no conflict of interest.

Author details

Farzaneh Abdolmaleki[1]*, Milad Daneshniya[1] and Mohammad Hossein Maleki[2]

1 Department of Food Science and Engineering, Faculty of Industrial and Mechanical Engineering, Qazvin Branch, Islamic Azad University, Qazvin, Iran

2 Department of Food Science and Technology, Science and Research Branch, Islamic Azad University, Tehran, Iran

*Address all correspondence to: fa.abdolmaleki@qiau.ac.ir

References

[1] Dewi T, Yustika RD, Arianti FD. Enhancement of production and food security through sustainable shallot cultivation. IOP Conference Series: Earth and Environmental Science. 2024: 1364. DOI: 10.1088/1755-1315/1364/1/012052

[2] Sonkamble SS, Kumbhar AC, Pawar KR, Nalawade SV. Sustainable cultivation of millets. International Journal of Multidisciplinary Research and Analysis. 2024;7:1971-1977. DOI: 10.47191/ijmra/v7-i05-20

[3] Umaharan P. Achieving sustainable cultivation of cocoa. Burleigh Dodds Series in Agricultural Science. 2018:590. DOI: 10.1201/9781351114547

[4] Hakmanage NM, Chandrasekara NV, Jayasundara DD. Comprehensive sustainable cultivation development plans based on statistical/mathematical models. International Journal of Scientific and Research Publications. 2020;**10**:255-259. DOI: 10.29322/ijsrp.10.08.2020.p10432

[5] Panda BP, Sharma SS. High value cropping—A small scale agro-business case from mushroom cultivation. Journal of Global Economy. 2018;**14**(4):1-9. DOI: 10.1956/jge.v14i4.492

[6] Higgins C, Margot H, Warnquist S, Obeysekare E, Mehta K. Mushroom cultivation in the developing world: A comparison of cultivation technologies. IEEE Global Humanitarian Technology Conference (GHTC). 2017;**2017**:1-7. DOI: 10.1109/GHTC.2017.8239314

[7] Pardo JE, Pardo-Giménez A, Zied DC, Álvarez-Ortí M. Evaluation of casing materials for mushroom production. Micologia Aplicada International. 2003; **15**(1):9-15. DOI: 10.5424/sjar/2008064-361

[8] Kokkora MI. Biowaste and vegetable waste compost application to agriculture [PhD thesis]. Cranfield, UK: Cranfield University, School of Applied Sciences, National Soil Resources Institute; 2008

[9] Kowalska AA, Grobelak A, Almås ÅR, Singh BR. Effect of biowastes on soil remediation, plant productivity and soil organic carbon sequestration: A review. Energies. 2020;**13**(21):5813. DOI: 10.3390/en13215813.

[10] Ravindran B, Karmegam N, Yuvaraj A, Thangaraj R, Chang SW, Zhang Z, et al. Cleaner production of agriculturally valuable benignant materials from industry-generated bio-wastes: A review. Bioresource Technology. 2020;**320**(Pt A):124281. DOI: 10.1016/j.biortech.2020.124281

[11] Soares AD, Prado RD, Bertani RM, da Silva AP, Deus AC, Kano C, et al. Contribution of using filter cake and vinasse as a source of nutrients for sustainable agriculture—A review. Sustainability. 2024;**16**(13):5411. DOI: 10.3390/su16135411

[12] Hardie M, Doyle R. Measuring soil salinity. Plant Salt Tolerance: Methods and Protocols. 2012;**913**:415-425. DOI: 10.1007/978-1-61779-986-0_28

[13] Khaniki GJ, Ghaderpoori M, Dehghani MH, Nazmara S. Analysis of toxic and trace metal contaminants in bottled water by using atomic absorption spectrometry. Food and Environment Safety Journal. 2011;**10**(2): 78-83

[14] Abdolmaleki F, Shirkhorshidi S, Daneshniya M. Investigating the characteristics of basil seed gum-based film enriched with Echinophora

platyloba extract and its preservative effect on the quality of silver carp. Research and Innovation in Food Science and Technology. 2023;**11**(4): 397-414. DOI: 10.22101/JRIFST.2022.321643.1320

[15] Xie Q, Li Y, Lv Z, Zhou H, Yang X, Chen J, et al. Effective adsorption and removal of phosphate from aqueous solutions and eutrophic water by Fe-based MOFs of MIL-101. Scientific Reports. 2017;**7**(1):3316. DOI: 10.1038/s41598-017-03526-x

[16] Banerjee P, Prasad B. Determination of concentration of total sodium and potassium in surface and ground water using a flame photometer. Applied Water Science. 2020;**10**(5):1-7. DOI: 10.1007/s13201-020-01188-1

[17] Vieira FR, de Andrade MC. Optimization of substrate preparation for oyster mushroom (*Pleurotus ostreatus*) cultivation by studying different raw materials and substrate preparation conditions (composting: Phases I and II). World Journal of Microbiology and Biotechnology. 2016; **32**:1-9. DOI: 10.1007/s11274-016-2152-y

[18] Van Loon PC, Swinkels HA, Van Griensven LJ. Dry matter content in mushrooms (Agaricus bisporus) as an indicator for mushroom quality. Mushroom Science. 2000;**15**:507-513

[19] Girmay Z, Gorems W, Birhanu G, Zewdie S. Growth and yield performance of Pleurotus ostreatus (Jacq. Fr.) Kumm (oyster mushroom) on different substrates. AMB Express. 2016; **6**:1-7

[20] Latifi L, Ghafouri Z, Manoochehri M, Khaki Arani M, Daneshniya M, Roohzadeh Nasiraei R, et al. Investigating the effect of using hydroxypropyl methylcellulose with watercress (Lepidium sativum) extract on reducing oil absorption and the quality of fried common carp (Cyprinus carpio) fillets. Journal of Food Science and Nutrition. 2021;**18**(3):125-139. Available from: https://sanad.iau.ir/fa/Article/832709

[21] Irshad A, Tahir A, Sharif S, Khalid A, Ali S, Naz A, et al. Determination of nutritional and biochemical composition of selected *Pleurotus* spps. BioMed Research International. 2023;**2023**(1): 8150909. DOI: 10.1155/2023/8150909

[22] Dehcheshmeh BK, Shakerian A, Rahimi E. Investigation of heavy metals, aflatoxin M1 and physicochemical properties of milk used in baby dry milk formula in Shahrekord City. Journal of Chemical Health Risks. 2022;**12**(3):527-535

[23] Roy S, Anantheswaran RC, Shenk JS, Westerhaus MO, Beelman RB. Determination of moisture content of mushrooms by Vis—NIR spectroscopy. Journal of the Science of Food and Agriculture. 1993;**63**(3):355-360. DOI: 10.1002/jsfa.2740630314

Chapter 5

The Comparison of the Vermicomposting of Cattle Dung and Leaf Litter as Sustainable Ways of Increasing the Nitrogen Concentration for Plant Nutrition and Soil Recovery

Alexander Kurovsky, Ekaterina Kornievskaya, Yana Gummer, Mohammed Sani Jibril, Israel Ogwuche Ogra and Andrey Babenko

Abstract

The average total nitrogen content in vermicomposts based on cattle dung, willow and poplar leaf litter was 2, 1.77 and 1.51%, respectively. The increase in total nitrogen content during vermicomposting of cattle dung, willow and poplar leaf litter was 40, 71 and 74%, respectively. Vermicompost based on cattle dung contained the highest nitrate nitrogen, 401 mg-eq/kg dry weight, while vermicompost based on willow and poplar leaf litter contained 207 and 126 mg-eq/kg, respectively. The actual activity of free-living nitrogen fixers is higher in vermicomposts based on woody leaf litter than in vermicomposts based on manure. The relationship between the increase in total nitrogen in the vermicompost and the actual activity of free-living nitrogen fixers was established. The number of nitrates in vermicomposts is related to the number of ammonifier microorganisms. The dynamics of nitrate release into the sandy substrate was faster for the vermicompost based on cattle dung and slower for the vermicompost based on poplar leaf litter. Vermicomposting of organic waste is a sustainable way of improving the Total Nitrogen for plant nutrition and soil recovery.

Keywords: cattle dung, willow leaf litter, poplar leaf litter, vermicompost, total nitrogen, nitrates, ammonifiers, free-living nitrogen fixers

1. Introduction

As a sustainable waste management method, vermicomposting uses earthworms and microorganisms to turn organic waste into nutrient-rich biofertilizers. There are substantial differences in the chemical content of vermicompost depending on the kind of feedstock, including manure, grass clippings, and leaf litter. The technique of vermicomposting involves using earthworms to break down organic waste, producing vermicast, a superior final product. Since vermicompost is nutrient-rich and acts as a soil conditioner, it is regarded as an organic fertilizer. The water-soluble nutrients in vermicompost improve soil structures like drainage and make additional types of nutrients available. Ingesting organic wastes including scarp papers, farmyard manure, crop residues, food and leftover residues, and yard trimmings, earthworms produce important goods like worm meal, vermicast tea, and worm casting, among others. Additionally, vermicompost combats a variety of plant ailments brought on by pests and soil-borne plant diseases by acting as an organic fertilizer and biological control agent. The use of vermicomposting can enhance other soil indicators, such as the nitrogen and C/N (carbon/nitrogen) ratio. Heavy metals' negative effects can also be lessened. Numerous beneficial microorganisms with a range of benefits are also provided by it.

Vermicomposting is a natural decomposition process where earthworms and microorganisms work together to mineralize organic waste substrates and transform them into nutrient-rich organic manure [1, 2]. It involves a bio-oxidative process in which earthworms work with the decomposer community (bacteria, fungus, and actinomycetes) to stabilize organic waste by changing its physical and biochemical characteristics [3]. The majority of the organic materials break down, while the leftovers are converted into stabilized vermicompost, which is high in humic compounds, hormones, and minerals [4]. Compared to chemical fertilizers, vermicompost possesses enormous benefits, especially for crops. Vermicompost is widely utilized as an organic fertilizer in organic farming on a big scale because it includes nutrients, humic acid, and growth hormones [5]. Several authors' studies have demonstrated that using vermicompost can improve crop output, vegetative development, and seed germination without sacrificing soil health. Fertility and soil health have been negatively impacted by intensive agriculture. By enhancing the physical, chemical, and biological characteristics of the soil, vermicompost application can aid in preserving its fertility and health [2, 4, 6].

The advantages of vermicomposting are numerous and include, a zero-waste and environmentally sustainable waste management technique, its applicability to nearly all kinds of nontoxic organic waste after preprocessing, the emission of fewer greenhouse gases and usage of less energy, inexpensiveness compared with conventional composting and its application in energy production and wastewater treatment [2, 7].

The fact that vermicompost is a fertilizer that contains all the necessary macronutrients for plants is now known to all those involved in agriculture in scientific, practical or household aspects.

We decided in this chapter to focus our attention only on nitrogen, among all the important nutritional elements because of the uniqueness of this macronutrient. Imagine a closed vermicomposting system where the process is carried out in containers that are isolated from the soil and interact only with atmospheric air (plastic containers, buckets, tanks). In this case, out of the triad of the most important macronutrients nitrogen, phosphorus, and potassium, the quantity of the last two in their total content will not change during vermicomposting. That is, the amount of phosphorus and potassium contained in the initial substrates will be the same as the number of phosphorus and potassium in the obtained vermicompost. And only the total nitrogen content can change (and upward) during vermicomposting. This is explained by microbiological mechanisms of nitrogen fixation, i.e., fixation of atmospheric nitrogen in soil (or soil-replacing substrates) through special groups of microorganisms. It is important to note that all the above-described peculiarities refer to the total forms of macronutrients, i.e., their potential total content, which is determined after the complete mineralization of the sample (e.g., using concentrated sulfuric acid). Thus, we are talking about potential reserves of elements for plant nutrition.

Another factor that motivates our special interest in nitrogen is that nitrogen nutrition is the most critical link in plant nutrition. The problem of plant uptake of nitrogen fertilizers in agriculture is quite acute. This is because the most oxidized form of mineral nitrogen – nitrates, is very poorly retained in the root-inhabited soil horizons and very easily washed out into deep soil horizons, inaccessible to plants. Therefore, it is always a problem to select fertilizers with controlled prolonged release of nitrogen in the rooting horizons. There is evidence that some organic fertilizers can release nitrogen more prolonged and optimally than mineral fertilizers.

Finally, the third reason for focusing on nitrogen is that it is the element that underpins the production of feed and food protein. In turn, humanity's global hunger problems are largely protein deficiency problems.

Turning to the types of vermicomposts, which we will write about in this chapter, we would like to note that in general vermicomposts can be divided into two groups according to the type of initial substrates – these are easily processed substrates rich in organics, including nitrogen-containing organics, and difficult-to-process substrates with a high content of lignin and cellulose. The most traditional is waste of the first group, called soft agriculture and farming waste. Soft agricultural waste refers to biodegradable organic materials obtained from agricultural processes that are easily decomposable. They are non-woody and contain high moisture content, making them good for vermicomposting. Examples include fruits and vegetable residues like peels and pulp, and other perishable organic matter generated through agricultural processing. Soft agricultural waste is easily degradable due to its lower lignin level and higher content of easily digestible compounds such as simple sugars [8]. This makes it specifically suitable for processes like composting and vermicomposting. The classic representatives of both groups are farm animal manure (for the first group) and woody leaf litter (for the second group).

Regarding the leaf litter, it should be noted that there is data on its use as a substrate for vermicomposting. However, in most cases, the fallow is used as an additional but not independent substrate for vermicomposting. Meanwhile, woody leaf litter has obvious advantages as a substrate for vermicomposting because it is produced in huge quantities in agroecosystems and urban ecosystems. The production of leaf litter from trees planted in public and private areas is a significant issue concerning their disposal. It is very tedious to collect them every day (or every season) and transfer them to the landfill location. In addition, the incorrect handling of these leaf litter at the disposal site pollutes the ecosystem. Dry leaves are occasionally burned, which is likewise an unethical method of cleaning up plant waste because it depletes the soil of vital nutrients and pollutes the air [9]. To maintain ecological balance and enhance natural resources, bio-organic waste must be managed effectively [10]. Various leaf litters, including wheat straw, sugarcane leaves, Ashoka tree leaves (*Polyalthia longifolia*), teak leaves (*Tectona grandis*), neem leaves (*Azadirachta indica*), tendu leaves, mango and guava leaves, rubber leaves, teak leaves, eucalyptus, black plum, and sandalwood, can be used to make vermicompost [11]. Dry leaf litter may be turned into a nutrient-rich substance and applied on agricultural land, making it a good and sustainable supply.

Trees of the willow family are widely distributed in the Northern Hemisphere in temperate regions. Furthermore, they are widely used in urban landscaping.

The following material will be presented on the peculiarities of the composting process of cattle dung and leaf litter concerning the nitrogen nutrition of plants.

2. Materials and methods

Experiments were conducted in different countries to process a wide variety of local organic wastes [12–17]. In the production of organic fertilizers, cattle manure is most commonly used. For more than 10 years our research group has been investigating the regularities of vermicomposting of woody leaf litter as a waste product of urban landscaping. We have always correlated the results obtained during the experiments with the results of vermicomposting of the traditionally widely used organic-rich substrate – manure (cattle manure, horse manure, etc.).

Experiments were conducted to compare vermicomposts based on manure with vermicomposts based on leaf litter by some physicochemical and microbiological indicators. The total nitrogen concentration and nitrate concentration were used as the main indicators of nitrogen reserves in vermicomposts. Microorganisms of trophic groups were used as microbiological indicators: ammonifiers, inorganic nitrogen utilizers, free-living nitrogen fixers, the number of which is counted on solid nutrient media.

Eisenia fetida (Savigny, 1826), one of the most widespread species of compost earthworms, was used for vermi-processing of substrates. This species is a cosmopolitan with a global distribution. It is relatively easy to grow in vermiculture at high densities and is tolerant to a fairly wide range of environmental changes (temperature 4–30°C, pH 5–9, humidity 60–90%). For vermicomposting we used the worm culture *Eisenia fetida,* maintained by the Head of the Department of Agricultural Biology of Tomsk State University, Professor Andrey Babenko.

Cattle dung (cow manure, cattle manure) was used in the experiments. Leaf litter was collected in the University Grove of Tomsk. For vermicomposting we used leaf litter of the two species: *Populus nigra* L. (black poplar), *Salix fragilis* L. (brittle willow). All the above-mentioned species grow in the same soil and climatic conditions. Consequently, all possible differences in physicochemical properties of leaf litter for these three tree species can be considered as genetically determined. The reason for choosing these two species of woody plants is described by us in [18].

The leaf litter was collected between 20 September and 20 October for 4 years, at a positive average daily temperature, until the formation of a stable snow cover. The collected leaf litter was dried at room temperature to constant weight and stored air-dry. Before vermicomposting, the litter was chopped to approximately 10 × 10 mm in size. Cow dung and leaf litter were food substrates for vermicomposting. Air-dry top peat was used as the bedding sorbent substrate.

Before vermicomposting, the food component and bedding substrate were mixed in a 1:1 mass ratio of 100 g and 100 g. Thus, the total mass of the dry mixture was 200 g; the volume occupied by each mixture was approximately 2 liters. The processed mixtures were moistened with distilled water 5 days before introducing the worms into the containers. The moisture content of the mixtures was brought to 75% by weight. By the time the worms were planted, the pH of the substrate was set at 5.5–5.7, which was acceptable for vermicomposting to begin. The worms and substrates were placed in plastic containers with ventilated lids with a volume of 7 liters. In each container with the prepared substrate, 20 *E. fetida* worms with a total weight of about 8 g were added. Vermicomposting took place in a dark room at an air temperature of +21 ± 3°C. The duration of vermicomposting was 3.5–4 months.

Compost option	Number of CFU in 1 g of vermicompost × 10^9		Actual activity of free-living nitrogen fixers, %
	Ammonifiers	Inorganic nitrogen utilizers	
Cow Manure vermicompost	7.85 ± 0.70	6.99 ± 0.64	76.48 ± 2.81
Poplar leaf litter vermicompost	5.02 ± 0.47	3.54 ± 0.57	89.59 ± 2.84
Willow leaf litter vermicompost	6.26 ± 0.82	5.77 ± 0.59	86.64 ± 2.27

Table 1.
Quantity of ammonifiers, inorganic nitrogen utilizers and free-living nitrogen fixers in vermicomposts.

At the end of processing, the substrates were freed from worms and cocoons. Before chemical analysis, vermicomposts were dried in a desiccator at 105°C.

The quantity of total nitrogen and nitrate in vermicomposts was determined using spectrophotometry and ionometry, respectively. For the determination of total nitrogen, the samples were subjected to mineralization in concentrated sulfuric acid in the presence of a selenium catalyst. Experiments to investigate the accumulation of nitrates leached from vermicompost into sandy substrate were conducted following the scheme "Soil Column Leaching Experiments" developed by us [19]. More details of our research methodologies can be found in our previous publications too [18, 20, 21].

The vermicompost was not dried before microbiological analyses. The number of ammonifiers and inorganic nitrogen utilizers was counted by limiting dilutions sown on meat-peptone agar and starch-ammonia agar, respectively, and expressed as the number of colony-forming units (CFU) per 1 g of absolutely dry compost. The actual activity of free-living nitrogen-fixing organisms of *Azotobacter* genus in percent was determined by the lump fouling method on Ashby's medium. All the microbiological analysis methods we used are described in detail in our previous papers [20, 21].

This chapter presents the minimal statistical treatment of the data summarized over several years, which includes the Kolmogorov–Smirnov test for normality hypothesis testing, calculation of means and 95% confidence intervals for the Student- and normal distribution. These two indicators (means and confidence intervals) represent the data in all figures and in **Table 1**.

3. Comparison of vermicomposting of cattle dung and leaf litter in the aspect of nitrogen nutrition of plants

Figures 1 and **2** show the content of total nitrogen and nitrate (respectively) in the initial substrates and vermicompost obtained from the three wastes.

The total nitrogen content in the initial substrate containing cattle manure was on average 1.43% (**Figure 1**). During vermicomposting of the manure, this indicator increased to 2%. The content of total nitrogen during vermicomposting of willow leaf litter increased from 1.06 to 1.81%. And at vermicomposting of poplar leaf litter - from 0.88 to 1.53% (**Figure 1**). Thus, the obtained vermicomposts were characterized by the content of total nitrogen in the range from 1.5 to 2%, which agrees well with

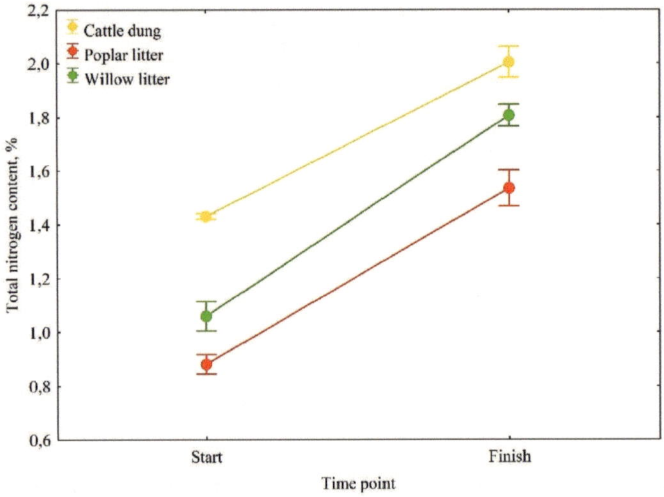

Figure 1.
The total nitrogen content in the beginning (hereinafter denoted as Start) and in the ending (hereinafter denoted as Finish) of the vermicomposting of cattle dung, willow- and poplar leaf litter.

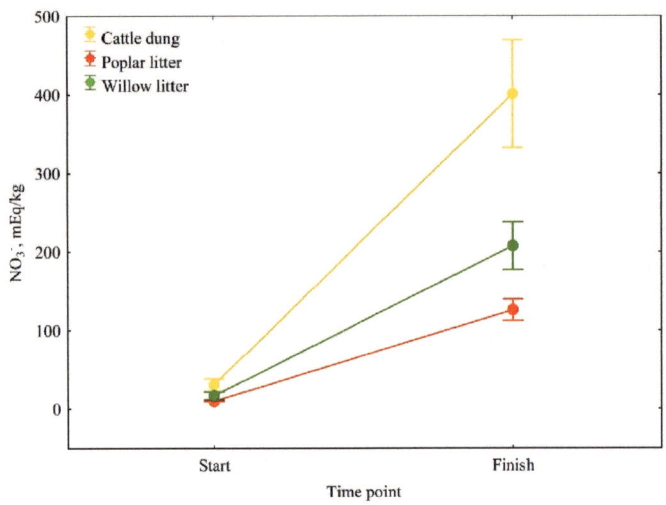

Figure 2.
The nitrate content in the beginning and in the ending of the vermicomposting of cattle dung, willow- and poplar leaf litter.

the literature data available to date [22]. The highest content of total nitrogen was in vermicompost based on cattle manure, and intermediate content of total nitrogen was recorded in vermicompost based on willow leaf litter, the lowest value of this indicator - in vermicompost based on poplar leaf litter, which, nevertheless, in terms of nitrogen content can also be considered as good organic fertilizer.

This means that the total nitrogen content in vermicomposts was always significantly higher than in the original substrates (**Figure 1**). The relative increase in total nitrogen content after vermicomposting for cattle manure was 40%, for willow fall 71% and for poplar leaf litter 74%. Thus, although the nitrogen content in vermicomposts based on woody leaf litter is lower than the nitrogen content in vermicomposts based on cattle dung, the relative increase in the content of this element in vermicomposting of litter is greater than in vermicomposting of dung.

The highest nitrate content in the obtained samples was in vermicompost based on cattle dung. The nitrate content was significantly lower in vermicomposts based on leaf litter from both tree species (**Figure 2**).

Special attention should be paid to the balance of total nitrogen during vermicomposting. For all three types of substrates, this balance was positive.

The nitrate content in vermicomposts based on willow and poplar leaf fall was 207 and 126 mg-equivalent, respectively (**Figure 2**). Thus, the nitrate content in vermicompost samples obtained by processing cattle manure was two or more times higher than that of vermicomposts obtained by processing leaf litter. As with the content of total nitrogen, the nitrate content increased many times during vermicomposting (**Figure 2**). Simple calculations show that to add an average number of 65 mg per kilogram of nitrate to any depleted soil, only about 15% (by weight) of manure-based vermicompost, about 30% of willow fall vermicompost and about 50% of poplar leaf litter vermicompost could be applied.

The findings during our research can be explained by using the data of microbiological analysis, which are presented below.

Ammonifiers and inorganic nitrogen utilizers were found in all tested vermicompost samples. Microorganisms of these groups were found in vermicomposts of about 109 CFU per 1 g of dry substrate. Free-living nitrogen-fixing organisms of the Azotobacter genus were also found in all vermicomposts. The number of microorganisms of these groups in different vermicomposts differed significantly (**Table 1**).

The data on the increase in total nitrogen (**Figure 1**) agreed well with the data on the actual activity of free-living nitrogen-fixing organisms in vermicomposts. In manure-based vermicompost, the actual activity of free-living nitrogen fixers was 76.48% and the total nitrogen content increased by 40%. In vermicompost based on willow fall these indices were 86.59 and 71% respectively, and in vermicompost based on poplar leaf litter the actual activity of free-living nitrogen fixers was almost 90% and the increase in total nitrogen was 74%.

It can be seen that both indices in the vermicompost series 'manure-based vermicompost – willow fall vermicompost – poplar fall vermicompost' increase. The correlation coefficient between the content of total nitrogen and the actual activity of free-living nitrogen fixers is close to 1, which demonstrates an unambiguous correspondence between these indicators of vermicomposts.

The data on nitrate content in vermicomposts agree well with such microbiological indicators as the amount of ammonifiers (**Table 1**).

Ammonia formed as a result of the ammonification process under aerobic conditions is inevitably involved in the nitrification process, which first produces nitrite and then nitrate, which is the most oxidized and plant-available form of nitrogen. Part of the ammonia produced in the ammonification process is used by inorganic nitrogen utilizers to increase their biomass. We can see that the number of inorganic nitrogen utilizers, as well as the number of ammonifiers, is minimal in vermicompost based on poplar leaf litter and maximum in vermicompost based on manure (**Table 1**). The activity of inorganic nitrogen utilizers can be seen as unfavorable for plants, as they convert nitrogen into a form inaccessible to plants. But, on the other hand, the conversion of inorganic nitrogen into organic nitrogen means that nitrogen in the soil or a soil-like substrate is stored in a reserve form which can be further converted by microorganisms into plant-available forms.

Thus, the activity of free-living nitrogen fixers significantly increases the total nitrogen content in vermicompost. Ammonifiers contribute to the conversion of nitrogen compounds into plant-available forms. Inorganic nitrogen utilizers contribute to the conservation of nitrogen in vermicompost. The number of microorganisms in these groups is related to the amount of nitrogen compounds in vermicompost. The results probably indicate that woody leaf litter decomposition in vermiculture is a more favorable substrate (in comparison to cattle dung) for the development of the activity of free-living nitrogen fixers.

Turning to the final part of our chapter, it should be said that the assessment of organic fertilizers only by their content or other elements of mineral nutrition of plants is not exhaustive. It is also very important to know how quickly or slowly a particular fertilizer releases certain nutrient elements. This information is particularly important for nitrogen.

Thus, according to the totality of the studied parameters, we identified two alternative vermicompost variants: vermicompost based on cattle manure and vermicompost based on poplar leaf litter. These two variants were tested for the character of the dynamics of nitrogen release (in the form of nitrate) from vermicompost into sandy substrate. We estimated the number of nitrates leached by water through the sand columns and converted this number per kg of the sand substrate (as model soil). **Figure 3** shows the results of experiments investigating nitrate leaching from two types of vermicomposts.

Figure 3 clearly shows that almost all the nitrate present in the vermicompost obtained from cattle manure processing was transferred to the sandy substrate during the first 3 days of the experiment. Thereafter, the nitrate content in the sand practically did not change. In contrast, elution of nitrates from vermicompost based on poplar leaf litter proceeded gradually during all 56 days of the experiment.

High content and almost instantaneous release of nitrates is a negative feature of the sample based on cattle manure because in the situation with real soil, excess nitrates are washed out into the underlying (inaccessible for plants) horizons, then into groundwater and may cause eutrophication of water bodies. The variant with

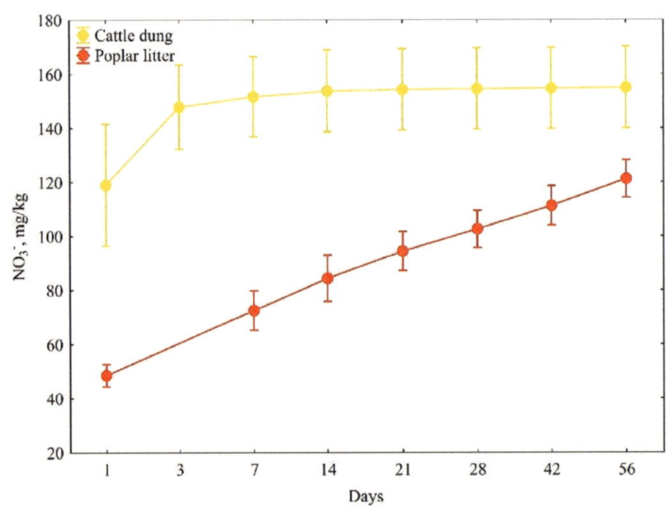

Figure 3.
The nitrate elution from the vermicompost based on cattle dung and poplar leaf litter into sandy substrate.

vermicompost based on black poplar leaf fall even after 2 months has the potential for further nitrate release into the soil.

4. Conclusion

Free-living nitrogen fixers, ammonifiers, and inorganic nitrogen utilizers were identified in vermicomposts. A reliable relationship between the increase in the number of total nitrogen and the actual activity of free-living nitrogen fixers was revealed. The amount of nitrates is related to the number of ammonifier microorganisms forming ammonia from organic nitrogen compounds. Vermicomposts contain a significant number of inorganic nitrogen utilizers that contribute to the preservation of nitrogen compounds in the substrate.

Each of the two types of substrates for vermicomposting (manure and woody leaf litter) has advantages and disadvantages. The advantage of vermicompost based on manure is the high content of both total nitrogen and nitrate. The disadvantage of the same vermicompost is a very high rate of nitrate return in the soil. That makes it somewhat similar to many mineral fertilizers. The leaf litter contains much less nitrogen in both forms, but the vermicompost that is obtained by its processing is more stable and has a potential for prolonged nitrate recovery, which is very important for the nitrogen nutrition of plants and the rehabilitation of depleted soils. In general, we can say that vermicomposting of organic waste is a way to sustainably increase nitrogen content in agroecosystems.

Author details

Alexander Kurovsky[1*], Ekaterina Kornievskaya[1], Yana Gummer[1], Mohammed Sani Jibril[1,2], Israel Ogwuche Ogra[1] and Andrey Babenko[1]

1 National Research Tomsk State University, Tomsk, Russian Federation

2 Department of Plant Biology, The Federal University Dutse, Nigeria

*Address all correspondence to: a.kurovskii@yandex.ru

References

[1] Pramanik P, Chung YR. Changes in fungal population of fly ash and vinasse mixture during vermicomposting by Eudrilus eugeniae and Eisenia fetida: Documentation of cellulase isozymes in vermicompost. Waste Management. 2011;**31**:1169-1175. DOI: 10.1016/j.wasman.2010.12.017

[2] Sharma K, Garg VK. Vermicomposting of waste: A zero-waste approach for waste management. Sustainable Resource Recovery and Zero Waste Approaches. 2019;**1**:133-164. DOI: 10.1016/B978-0-444-64200-4.00010-4

[3] Gomez-Brandon M, Dominguez J. Recycling of solid organic wastes through vermicomposting: Microbial community changes throughout the process and use of vermicompost as a soil amendment. Critical Reviews in Environmental Science and Technology. 2014;**44**(12):1289-1312. DOI: 10.1080/10643389.2013.763588

[4] Hussain N, Abbasi SA. Efficacy of the vermicomposts of different organic wastes as "clean" fertilizers: State-of-the-art. Sustainability. 2018;**10**(4):1205. DOI: 10.3390/su10041205, 10.1080/10643389.2013.763588

[5] Dominguez J, Edwards CA. Biology and ecology of earthworm species used for vermicomposting. In: Vermiculture Technology: Earthworms, Organic Wastes, and Environmental Management. Boca Raton, FL: CRC Press; 2011. pp. 27-40

[6] Chowdhury A, Sarkar A. Vermicomposting—The sustainable solid waste management. Waste Management and Resource Recycling in the Developing World. 2022;**1**:701-719. DOI: 10.1016/B978-0-323-90463-6.00013-0

[7] Sharma K, Garg VK. Comparative analysis of vermicompost quality produced from rice straw and paper waste employing earthworm *Eisenia fetida* (Sav.). Bioresource Technology. 2018;**24**(8):7829-7836. DOI: 10.1016/j.biortech.2017.11.101

[8] Cazier EA, Pham T, Cossus L, et al. Exploring industrial lignocellulosic waste: Sources, types, and potential as high-value molecules. Waste Management. 2024;**188**:11-38. DOI: 10.1016/j.wasman.2024.07.029

[9] Porichha GK, Hu Y, Rao KT, et al. Crop residue management in India: Stubble burning vs. other utilizations including bioenergy. Energies. 2021;**14**(14):4281. DOI: 10.3390/en14144281

[10] Narula GB, Soni K, Vadakan AB, et al. Bio-organic approach for monitoring the plant growth for sustainable agriculture. Journal of Agriculture and Applied Biology. 2023;**4**(2):161-173. DOI: 10.11594/jaab.04.02.06

[11] Gnanasekaran R, Jerin RS, Dhanalakshmi M, et al. Utilization of Lagerstroemia speciosa dry leaf litter combined with cattle dung for the production of enriched vermicompost–a possibility of valorization. International Journal of Recycling of Organic Waste in Agriculture. 2023;**12**(4):699-708. DOI: 10.30486/IJROWA.2023.1977455.1580

[12] Jayakumar P, Natarajan S. Microbial diversity of vermicompost bacteria that exhibit useful agricultural

traits and waste management potential. Springerplus. 2013;**1**:1-26. DOI: 10.1186/2193-1801-1-26

[13] Karthikeyan M, Gajalakshmi S, Abbasi SA. Effect of storage on the properties of vermicompost generated from paper waste: With focus on pre-drying and extent of sealing. International Journal of Energy and Environmental Engineering. 2014;**5**:291-301. DOI: 0.1007/s40095-014-0135-z

[14] Emperor GN, Kumar K. Microbial population and activity on Vermicompost of Eudrilus eugeniae and Eisenia fetida in different concentrations of tea waste with cow dung and kitchen waste mixture. International Journal of Current Microbiology and Applied Sciences. 2015;**4**(10):496-507

[15] Chen Y, Chang SKC, Chen J, et al. Characterization of microbial community succession during vermicomposting of medicinal herbal residues. Bioresource Technology. 2018;**249**:542-549. DOI: 10.1016/j.biortech.2017.10.021

[16] Gopalakrishnan S, Vadlamudi S, Bandikinda P, et al. Evaluation of Streptomyces strains isolated from herbal vermicompost for their plant growth-promotion traits in rice. Microbiological Research. 2014;**169**:40-48. DOI: 10.1016/j.micres.2013.09.008

[17] Nweke IA. Plant nutrient release composition in vermicompost as influenced by Eudrilus eugenae using different organic diets. Journal of Ecology and the Natural Environment. 2013;**5**(11):346-351. DOI: 10.5897/JENE2013.0401

[18] Kurovsky AV, Petrochenko KA, Godymchuk AY. Physicochemical aspects of recycling tree leaf litter in the south of Western Siberia by the Eisenia fetida (Savigny) vermiculture. IOP Conference Series: Earth and Environmental Science. 2019;**226**:012009. DOI: 10.1088/1755-1315/226/1/012009

[19] Rudmin M, Banerjee S, Makarov B, et al. Glauconite-urea nanocomposites as polyfunctional controlled-release fertilizers. Journal of Soil Science and Plant Nutrition. 2022:4035-4046. DOI: 10.1007/s42729-022-01006-4

[20] Kornievskaya EV, Kurovsky AV, Babenko AS, et al. Microbial structure of nitrogen utilizers in *Populus nigra* L. compost and vermicompost. IOP Conference Series: Earth and Environmental Science. 2020;**433**:012001. DOI: 10.1088/1755-1315/433/1/012001

[21] Kurovsky AV, Kornievskaya EV, Gummer YM, et al. The balance of nitrogen forms and number of microorganisms of the nitrogen cycle in vermicomposts based on leaf litter and cow manure. IOP Conference Series: Earth and Environmental Science. 2021;**935**(1):012002. DOI: 10.1088/1755-1315/935/1/012002

[22] Adhikary S. Vermicompost, the story of organic gold: A review. Agricultural Sciences. 2012;**3**:905-917

Chapter 6

Bee Propolis as a Biological Alternative to Control Fungi in Lemon Cultivation

José Alfonso López García, Julieta Grajales Conesa, Víctor Jesús Albores Flores, Maria Guadalupe De Gyves Córdova, Rodolfo Torres de los Santos, Hermilo Lucio Castillo, Marycarmen Utrilla Vázquez, Betsabé Rodríguez Pérez and Leopoldo Caridad Cruz López

Abstract

In this chapter, an alternative for damage control of lemon trees in the field and its monitoring until the post-harvest stage is presented, with the use of propolis extracts, applied by spraying to the trees and placed in biofilms in fruits after harvest. The objective of this work was to evaluate the biological effectiveness of ethanolic extracts of propolis from *Scaptotrigona mexicana* and *Melipona beecheii* under field conditions and in the post-harvest stage in Mexican lemon (*Citrus aurantifolia*) for the production and conservation of quality in the state of Chiapas, Mexico. A 70% decrease in foliar damage was found in the leaves of lemon trees, that the most resistant area of the leaf is the one attached to the petiole. The chitosan biofilms with propolis allowed the shelf life of the fruits to last 15 days, allowing the variables of color, weight loss, firmness, phenols, and flavonoids to be preserved in permissible ranges. The above allows us to propose propolis extracts as another alternative in phytosanitary management programs in the agricultural production of lemon crops.

Keywords: propolis, lemon, health, biofilms, alternative

1. Introduction

Until 2020, Mexico has occupied on average 207 thousand hectares of lemon cultivation area and produces an average volume of 2.8 million tons, according to data published on the portal of the Agri-Food and Fisheries Information Service [1]. Currently in Mexico, there are records of around 50 pathogens that cause damage to citrus; some of them are of fungal origin of economic importance, producing diseases, such as *Phytophthora citrophthora* (gummosis), *Colletotrichum acutatum* (anthracnose), *Capnodrium citri* (black damping-off), *Mycosphaerella citri* (greasy spot), *Elsinoe fawcettii* (scab), and citrus leaf spot (*Pseudocercospora angolensis*), which is a fungus that requires moisture for its spread and infection, produces circular chlorotic spots of 4–10 mm in diameter on the leaves. Due to the environmental conditions present in Soconusco, Chiapas, the proliferation and infection of these microorganisms affect lemon cultivation [2, 3].

On the other hand, in the post-harvest stage, the problem that lemon presents is that it has a very accelerated life period; its useful life is low in environmental conditions, it is 10–15 days, causing a decrease in quality standards [4, 5]. Lemon has bioactive compounds, such as vitamins, carotenoids, fiber, anthocyanins, phenolic compounds, and flavonoids, the latter two being of great importance because they are responsible for providing beneficial properties such as antioxidants, which prevent the negative activity of free radicals [5, 6]. The degree of ripeness of the lemon not only affects the commercial quality but also gradually decreases the biological activity [7]. To preserve the quality of the fruits, various physical techniques have been created, such as freezing, which causes irreversible changes in the product, heat treatments causing losses of nutrients, decrease in weight, flavor, and vitamins. Chemical treatments for quality preservation can cause health problems for consumers due to their lack of regulations and indiscriminate use, but the application of edible coatings may be a more viable option to address the shortcomings of these methods [8].

Among the biological control alternatives, hive products, such as propolis, have been observed as an option for the management of citrus crops, both at the field level and post-harvest. Propolis is a substance composed of resins from the surrounding vegetation, balms (50%), waxes (30%), aromatic and essential oils (10%), pollen (5%), and other substances (5%). Its composition varies due to intrinsic factors such as the different species of bees as well as genetic differences between the same species; likewise, there are extrinsic factors, such as humidity, temperature, the collection area, and the type of plants from which the materials with which this substance is manufactured are obtained [9–11].

Edible coatings, as a method of fruit preservation, help to prolong shelf life, delay decomposition, maintain organoleptic and nutritional properties in optimal conditions, and preserve quality [8, 12]. These edible coatings can be defined as a transparent and thin matrix, free of taste and odor, because they are made of biopolymers [13–15]. Chitosan is a polysaccharide widely used in biofilm formulation due to its easy processing, low cost, and easy handling, and can be combined with other compounds, such as propolis, with the aim of prolonging the shelf life of fruits [16].

The use and application of propolis, with and without a matrix that keeps it fixed (such as biofilms), have been shown to present antifungal activity against a variety of plant pathogenic fungi at both the field and post-harvest stages [17, 18]. In addition, as a component of the protective cover of fruits, it supports increasing beneficial properties, such as improving the firmness of the fruits and reducing weight loss, among others [19–22]. The production of biofilms based on propolis focuses on the use of *Apis mellifera* because it helps preserve the quality of various fruits, delays their decomposition, and creates an antioxidant barrier thanks to its antioxidant biological property [23]. Stingless bees also have the same properties, which indicates that they can be used in the formulation of biofilms [9].

The objectives involved for the present study were to evaluate the biological effectiveness of ethanolic extracts of propolis from *Scaptotrigona mexicana* and *Melipona beecheii* against citrus leaf spot in the cultivation of Creole lemon (*Citrus aurantifolia*) at the field level in the state of Chiapas, Mexico, and the second was to determine the effectiveness of chitosan biofilms added with propolis from stingless bees in Mexican lemon (*Citrus aurantifolia*) for quality conservation.

2. Materials and methods

In the field stage, propolis obtained from the Association of Meliponicultors of Soconusco S. C. del R. L. was used, which includes the municipalities of Tapachula, Tuxtla-chico, Unión Juárez, Izapa, Francisco I. Madero, Cacahoatán, and Mazatán in the region. Soconusco, Chiapas. The stingless bee species were the following: *S. mexicana* and *M. beecheii*. The samples were collected with previously sterilized stainless-steel tools, taking approximately 1.5 kg of propolis. The samples were transferred to the laboratory and kept refrigerated at 4–8°C.

Ethanolic extracts were prepared at a concentration of 25% w/v of both species *S. Mexicana (S. m.)* and *M. beecheii* (M. b.). A total of 50 g of crushed and homogenized propolis were weighed and placed in sterile bottles protected from light with 100 mL of 96% ethanol, kept stirring manually for 10 minutes for 7 days at room temperature. Subsequently, the contents of the bottles were filtered through Whatman No.1 paper, the residues were washed with ethanol twice with 50 mL, bringing the final sample volume to 200 mL, and the extract was kept refrigerated at 4–8°C in amber bottles in the absence of light [24].

In the field, a concentration of 25% was used, conditioning the age of the lemon tree, and at the post-harvest level, the concentrations were 5, 10, and 15%, all percentages are w/v. Field stage. The study was carried out in a 4-year-old commercial lemon plantation in the Las Varillas Canton, in the municipality of Mazatán, Chiapas, Mexico (Lat 14°55′17" N, Lon 92°31′54" W).

Ten plants were selected for each treatment for a total of 80 plants, using a completely randomized experimental design with a strip arrangement with 10 repetitions and five treatments plus a positive control and a negative control in adult and juvenile trees. Ethanolic extracts were used at a concentration of 25% w/v of both species *S. mexican* and *M. beecheii*, to form the following treatments: T1- *S. mexican* propolis at 25% w/v applied to adult trees; T2- *S. mexican* propolis at 25% w/v applied to juvenile trees; T3- *M. beecheii* propolis, at 25% w/v applied to adult trees; T4- *M. beecheii* propolis, at 25% w/v applied to juvenile trees; T5- Combination of *S. mexican* and *M. beecheii* propolis at 25% in juvenile trees; Control - Copper oxychloride + mancozeb in adult and juvenile trees.

2.1 Application of treatments in the field

A total of 200 mL of the ethanolic extracts were dissolved in a final volume of 15 L. Using as equipment for the application a backpack-type hand sprayer pump with a capacity of 20 L Brand Pretul Model: FUM-20P, a total of five applications every 14 days, applying the treatments to all the foliage, on the upper and lower sides of the leaves, floral panicles, and fruits, in the morning hours and in the absence of rain, different colored tapes were used to distinguish the treatments [25].

2.2 Variable evaluation

The data recording was carried out in the following way: each plant was divided into four quadrants according to the cardinal points, selecting a branch with healthy vegetative shoots per quadrant, and carrying out evaluations every 7 days identifying symptoms of anthracnose infection produced by *Colletotrichum ssp*, which consist of circular lesions with a colored center gray to black, appearing simply or surrounded by a yellow halo with a chlorotic appearance [26].

2.3 Severity

To determine the severity of the damage, the leaf was divided in half, and the number of leaves with damage in the apex area, damage in the petiole area and damage in the apex and petiole was counted on each branch of the selected shoots. The criteria considered were grade zero (no symptoms), grade 1 (damage to the apex), grade 2 (damage to the petiole), and grade 3 (damage to the apex and petiole).

2.4 Post-harvest stage

2.4.1 Preparation of chitosan and preparation of biofilms

Chitosan (Encapsulators) was used (15 gL-1), glacial acetic acid (Binden Chemical), (10 mL L-1), and glycerol (Meyer) (1 mL L-1). The chitosan solution was prepared from a concentration of 1.5% with a volume of 200 mL, kept under constant stirring for 24 hours at 30°C, and the pH was adjusted to 4.6 [27].

2.4.2 Biofilm preparation

Propolis extract from *Scaptotrigona mexicana* and *Melipona beecheii* was added to the chitosan solution. It was added slowly and kept under constant stirring for 30 minutes, then Tween 20 (Meyer) was added (10 mL L-1) to form an emulsion and heated at 40°C on a grill for 10 minutes [27, 28].

2.4.3 Treatments

Eight treatments were carried out, fruits without treatment were called the negative control, and fruits with chitosan were called the positive control. The treatments are described below: 1 (S. m. 5%), 2 (S. m. 10%), 3 (S. m. 15%), 4 (M. B 5%), 5 (M. b. 10%), 6 (M. b. 15%), 7 (Chitosan 1.5%), 8 (control).

2.4.4 Application of treatments

Twenty-four fruits were used for each treatment. The fruits were washed and disinfected with 3% sodium hypochlorite; the treatments were applied by immersion for 30 seconds, allowed to dry at room temperature, and stored for 15 days [28].

2.4.5 Evaluations

The incidence was determined by dividing the healthy fruits by the diseased ones, multiplying by 100.

Weight loss: The weight of the fruits was recorded using a digital scale. The fruits were weighed every third day, and the results were expressed in percentage of weight loss [28].

Firmness: The firmness of the fruit was analyzed using a manual 8 mm diameter cylindrical tip penetrometer (SUNDOO MODEL GY-1). The firmness was taken every third day, and the fruits were expressed in N.

Total phenols and total flavonoids: To obtain phenols, the lemon peel was macerated in alcohol at 70° for 1 day. A gallic acid calibration curve was carried out from a 100 mL stock solution at a concentration of 100 mg/mL; from six different concentrations were prepared from this solution—0, 10, 25, 50, 75, and 100 mg/mL [29] with modifications. The determination of phenols was done using the Folin–Ciocalteu colorimetric method, using the Folin–Ciocalteu reagent 1:10, 7.5% sodium carbonate, and 3 mL distilled water [28] with modifications. All absorbances were read at 740 nm in the spectrophotometer. The results were expressed in milligram equivalents of gallic acid per gram: (mg EAG/g). *Total flavonoids*: To obtain flavonoids, the lemon peel was macerated in 70° alcohol for 1 day [29]. A quercetin calibration curve was carried out from a 100-mL stock solution at a concentration of 5 mg/mL. From this solution, six different concentrations were prepared: 0.0, 0.3, 0.5, 1.0, 2.5, and 3.0 mg/mL, [30] with modifications. The determination of flavonoids was done using the aluminum trichloride method, using the 2% aluminum trichloride reagent and 3 mL distilled water [30] with modifications. All absorbances were read at 490 nm in the spectrophotometer. The results were expressed in milligram equivalents of quercetin per gram (mg EQ/g).

All the data obtained were evaluated to determine ANOVA and AMOVA (the latter where it could be carried out), as well as a comparison of means by Tukey and by Hotelling and the Pearson correlation. A principal component analysis was also performed. The software used was Infostat 2022.

3. Results

3.1 Field stage

Table 1 and **Figure 1**, show the dynamics presented by the lemon trees treated with Fungicide, *M. beecheii* propolis and *S. mexicana*. Highly significant differences were found between treatments and between evaluations ($p < 0.0001$). It is

	Melipona becheii				*Scaptotrigona mexicana*				Control			
E	**Ls**	**Ld**	**LdA**	**Cdl**	**Ls**	**Ld**	**LdA**	**Cdl**	**Ls**	**Ld**	**LdA**	**Cdl**
1	2	0.3	0.34	1.1	1.4	0.2	0.25	1	3.75	1.53	1.07	1.19
2	2	0.27	0.27	2.1	1.6	0.27	0.27	1.6	1.6	0.44	0.45	0.69
3	3.45	1.3	1	1	3	1.1	0.8	1	2.81	0.76	0.83	1.75
4	1.8	0.44	0.45	0.69	1.3	0.44	0.45	0.69	2.74	0.62	0.53	1.65
5	1.6	0.4	0.52	0.6	1.3	0.2	0.4	0.5	1.89	0.56	0.42	0.8
6	1.5	0.56	0.36	0.6	1.1	0.4	0.3	0.5	1.68	0.45	0.52	0.76
7	2	0.5	0.53	1.3	1.4	0.4	0.45	0.8	2.1	0.4	0.47	1.53
8	2.1	0.5	0.63	1.3	1.5	0.5	0.5	0.9	2.63	0.43	0.39	2.12

Table 1.
Damage dynamics in leaves of lemon trees in all treatments (E: Evaluation, Ls: Leaves with signs, Ld: Leaves with damage, LdA: Leaves with damage at the apex, Cdl: Completely damaged leaves).

Figure 1.
Lemon tree leaves before (A) and after (B) the application of propolis treatments.

observed that at the beginning, the severity of leaves with the presence of signs of the disease was greater in the control treatment, observing a reduction in this value in the second evaluation. All treatments presented an increase in sign values on the leaf in the third evaluation. Similarity in values was observed between the control and *M. beecheii*, with *S. Mexican* being different, where the number of leaves with signs was lower.

Regarding damaged leaves, on the equatorial part, the control treatment and that of *S. mexicana* presented lower values, compared to that of *M. beecheii*. In leaves that began with damage at the apex, only the *S. mexican* treatment presented the lowest values from the second evaluation, maintaining this dynamic throughout the period of time that the study was carried out.

3.2 Post-harvest stage

The difference between treatments was highly significant in phenolic compounds ($p < 0.0001$). With the exception of the treatment that included only chitosan biofilm (**Table 2**), which showed an increasing concentration of phenolic compounds, all other treatments showed a dynamic of maintaining the concentration of phenolic compounds. A decrease in this concentration was observed on days 6 and 9, with a subsequent increase on the remaining days. Only the treatment of *S. mexicana* with 10% propolis showed an increase in phenols on day 6, decreasing on day 9. Only 15% *M. beecheii* propolis treatment, starting on day 9, showed a similar behavior to the chitosan-only treatment.

Regarding the content of flavonoid compounds (**Table 2**), it was observed that in the treatments that did not receive a percentage of propolis solution, considered as control treatments, the dynamics decreased until day 9, where it subsequently increased and was located between concentrations similar to the other treatments. All other treatments showed similar dynamics, observing that on days 3, 9, and 12, some treatments presented high flavonoid values. All other treatments showed similar dynamics, with some treatments showing high flavonoid levels on days 3, 9, and 12. The observed range of flavonoid concentrations was generally between 0.18 and 0.3 mEQ/g. The difference between treatments was highly significant in flavonoid concentration ($p < 0.0001$).

T	C	Phenols (mg EAG/g)						Flavonoids (mg EQ/g)					
		Days						Days					
		0	3	6	9	12	15	0	3	6	9	12	15
BPSM	0	39.42	34.31	30 42	29.44	37.48	37.17	0.19	0.207	0.19	0.17	0.22	0.26
BPSM	5	38.70	36.80	31.83	29.81	35.94	27.86	0.17	0.23	0.21	0.20	0.21	0.25
BPSM	10	38.52	34.6	40.92	31.78	39.71	37.80	0.19	0.22	0.23	0.22	0.25	0.25
BPSM	15	37.41	38.13	29.52	31.13	37.25	38.33	0.202	0.21	0.22	0.21	0.26	0.25
BPMB	0	39.42	34.61	31.42	29.66	36.43	37.36	0.19	0.21	0.18	0.18	0.23	0.25
BPMB	5	34.00	33.38	28.24	34.10	40.51	36.95	0.19	0.21	0.18	0.24	0.24	0.24
BPMB	10	34.85	33.94	30.86	34.55	24.29	38.31	0.19	0.25	0.19	0.25	0.23	0.25
BPMB	15	36.65	29.71	25.36	42.77	41.18	37.86	0.20	0.20	0.193	0.27	0.23	0.25
BSPQ	1.5	22.33	27.86	28.4	43.26	41.02	36.96	0.20	0.20	0.18	0.28	0.23	0.26

Table 2.
Dynamics of phenolic compounds and flavonoids in all treatments (T: treatment, C: concentration).

In all treatments corresponding to propolis, the loss of firmness in the fruits was observed on day 6 onward (**Figure 2**), unlike the control that received only chitosan. With chitosan-only treatment, firmness loss was already observed from day 3, being consecutive until reaching a value of 28 N/cm^2. In the 5, 10, and 15% propolis treatments, regardless of the bee species, the dynamics of loss of firmness were not so steep, achieving a constant loss in this variable. No statistical differences were found between treatments.

In **Figure 3**, you can see the dynamics of weight loss, which was very constant in all treatments. Treatment of *M. beecheii* with 15% propolis, weight loss decreased after the sixth day, observing a very slight amount of this value in the remaining days of the study. Something similar was observed in the treatment of *S. mexicana* with 10% propolis, differentiating after day 12 of the study, where a considerable weight

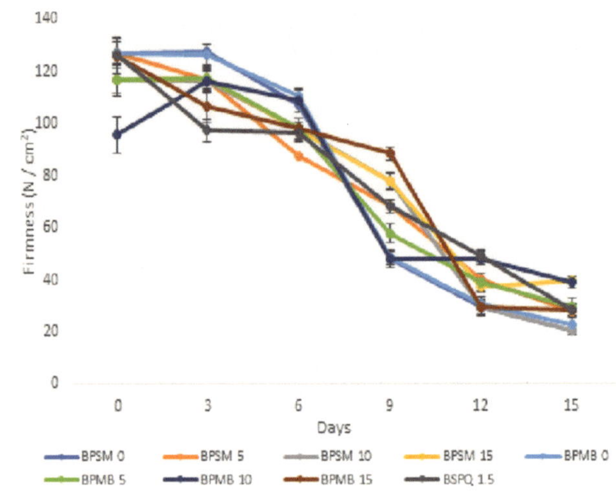

Figure 2.
Dynamics of firmness change of lemon fruits in all treatments.

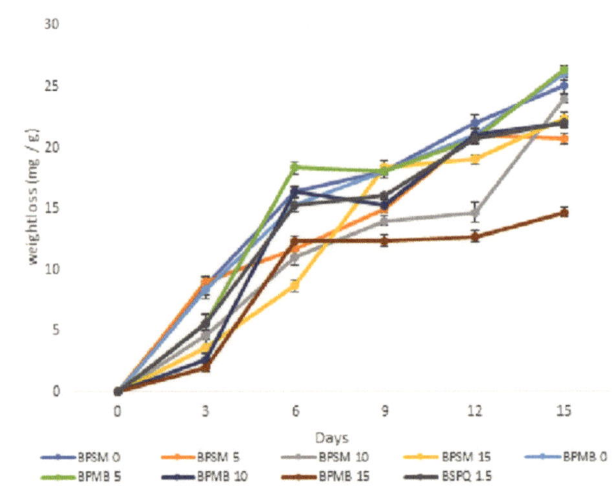

Figure 3.
Weight loss dynamics of lemon fruits in all treatments.

	Treatments								
	BPSM	BPSM	BPSM	BPSM	BPMB	BPMB	BPMB	BPMB	BSPQ
	0	5	10	15	0	5	10	15	1.5
Incidence	66	60	52	43	68	63	55	45	65

Table 3.
Incidence of lemon fruits by treatment (The numbers 0, 5, 10, and 15, are concentration value).

loss value was again presented. The statistical differences between treatments were highly significant ($p < 0.0001$).

The treatment of *S. mexican* with 15% propolis stopped losing considerable weight from day 9 of the study, with small values being observed later. In the rest of the treatments, starting on day 6, a reduction in weight loss was observed without presenting small values that would show a significant reduction. The observed dynamics of these treatments, after day 6, allowed us to see a consecutive line with values higher than the *M. beecheii* treatments of 10% and *S. mexicana* of 10 and 15% of propolis, presented in the remaining days of the study.

The percentage of diseased fruits was inversely related to the concentration of propolis applied (**Table 3**). This effect was observed to be greater with the species of *S. mexicana*, compared to that of *M. beecheii*. The results found in the chitosan biofilm were similar to those in treatments where propolis was not applied. The differences were highly significant between treatments ($p < 0.0001$).

4. Discussion

4.1 On field

Since 2000, several studies have been carried out on the control of pathogens in general, which have shown that propolis extracts can be used not only in the medical field. Recent studies, no more than 15 years ago, have shown that the use of propolis via ethanolic propolis extract (EEP) provides beneficial effects on crops, reducing the populations of phytopathogens associated with agricultural diseases, post-harvest preservation and conservation of fruits and vegetables, and promoting plant growth. These advances have shown that the chemical composition of each propolis, depending on the place or region, its action lies in the type, variety, and quantity of phenolic compounds and nutrients that constitute it [31]. These ethanolic extracts of propolis (EEP) do not affect the organs of plants and trees, as demonstrated in the present study. The results obtained in lemon trees by applying propolis from *Melipona becheei* and *Scaptotrigona mexicana* in the field demonstrate that it is a product of the hive, does not damage any organ of the trees, and that fungal diseases can be controlled with foliar applications.

The significant reduction of signs and symptoms observed in the leaves of lemon trees treated with EEP, regardless of the concentration used, suggests that: 1. the constituent components of EEP, at all concentrations used, presented antifungal activity, as indicated by Cibanal [32]. 2. The extraction carried out in the preparation of EEP is viable and does not require another process to extract secondary metabolites with this activity, as indicated by Retamoso [33] and Joya [34]. 3. The lack of symptoms and signs in new lemon leaves is related to the antimicrobial activity of the phenolic and flavonoid compounds present in EEP. Retamoso [33] indicating that these compounds present bioactivity and are compatible with plants, and Romero [35], this bioactivity will depend on the location, botanical species, and region. 4. Suggests that these compounds penetrated the cuticle and acted as eliciting agents that increased the defenses of lemon trees. The penetration of the constituent compounds of the EEP through the plant cuticle was due to the electrochemical-polar interaction and the hydrophobic properties that these constituents present [36], which facilitated migration through the lipid layers of the cuticle [37]. Phenolic compounds have been characterized by presenting a variety of biological activities, among which the importance of being an elicitor of defense mechanisms in plants has regained, which increases protection and improves the health of all plant organs [38]. Therefore, EEPs can be used to induce defense mechanisms in plants.

The damage observed in different parts of the leaf, such as close to the petiole, in the equatorial zone, or at the apex of the leaf. In addition to the properties of propolis, the physical properties were considered in the explanation, such as the shape of the leaf, the inclination of the leaf, the viscosity of the EEP solution, the air, the ambient temperature of the ground, and the interactions of the components of the solution with the surface of the leaf (which is hydrophobic). The above suggests that the viscosity of the EEP solution, which falls on the surface tension of each layer of liquid, promoted the retention of the solution in the leaf area (called retention time), thereby causing a differential location of damage on the leaf of lemon trees (**Table 1**). The retention time of a hydrophobic molecule on a hydrophobic layer increases the permeability of the compound through it, affecting the (minimum) retention and allowing its migration to the other end of the layer (this phenomenon is independent of the concentration of the compound) [39].

4.2 In post-harvest

The variation of phenolic compounds in the first days of fruit storage, only in propolis treatments, are related to the metabolic processes of production and antioxidant activity, reducing phenols in the first 6 days due to their activity and the generation of other compounds after day 6, essential in the phenological changes of the fruit during storage. In our present study, it is emphasized that lemon fruits were not refrigerated and remained at a temperature that varied from 28 to 36 degrees Celsius, receiving heat from outside, which caused constant changes. Heat as an inducer of changes in fruit was studied by Brizzolara [40], who studied the effects of temperature and determined that they alter the physiology and biochemistry of the fruit, which produces changes in the composition of metabolisms and compounds related to carbon and nitrogen, all this during the days in which it remains in storage, with and without temperature control. This was similar to what Amodio [41] indicated when observing it in purslane, apple, broccoli, lemon, and mandarin, Oliveira [42] in strawberry fruits and Martinez [43] in strawberry fruits, too.

Unlike the propolis treatments, in the treatment that only received the chitosan biofilm, the dynamics of phenolic compounds was on an increasing trend, reaching a maximum point on day 9 and then decreasing, until finally presenting concentrations similar to those of the other treatments. This suggests that chitosan alone is a component that induces the production of phenolic compounds, called elicitor, and has promoted the fruit's defenses. Anaya-Esparza [44] and Xoca-orozco [45], indicated that the interaction of chitosan with the pericarp of the fruits induces the expression of genes related to the biosynthesis of phenolic compounds and regulates a large number of metabolic processes that prevent the proliferation of phytopathogenic microorganisms.

All treatments, except for the 0% propolis treatments, showed an induction of phenolic compounds and flavonoids (**Table 2**). The latter are strongly attributed to antimicrobial activity, and their concentration was approximately above 0.21 mgEQ/g until day 9 and on day 12 above the value of 0.23 mgEQ/g. The increasing dynamics of these compounds in most of the treatments with propolis and chitosan suggest a protection of lemon fruits, but on the other hand, in **Table 3**, it was observed that the treatments with the different concentrations of propolis presented a lower incidence value with a Pearson correlation with the flavonoids of 0.68 with *M. beecheii* propolis, and 0.72 with *S. mexicana* propolis. These results suggest that protection was greater with propolis treatments and that the phenolic and flavonoid compounds generated in these treatments presented greater antifungal activity than those produced with chitosan treatment. These results agree with those reported by Mulkay [46], who evaluated red and brown propolis in mango fruits at the post-harvest stage, indicating that 35% less incidence and 40% less severity of anthracnose were achieved. Anaya-Esparza [44] agrees with Mulkay [46] by indicating that greater healthiness increases the shelf days of the fruits and, in parallel, improves the quality indicators of the fruits, such as firmness, pH, weight loss, and acidity and brix degrees.

The behavior of the control treatments forms a sigmoid in a decreasing sense of the loss of firmness of the lemon fruits, unlike the other treatments (**Figure 2**). The chitosan treatment presented a constant of this variable between days 3 and 6 and then maintained a decreasing line. This could be related to a reduction in metabolic activity, which corresponds to changes in biochemical activity (molecules) and physiological activity, which imply cellular changes and oxidative-reductive processes such as respiration, which between days 3 and 6 decreased their speed, as observed by Martínez-Gonzáles [43].

These changes (**Figure 2**), to a lesser extent, were observed in the propolis treatments between days 0 and 6, where the values of firmness were lower than in the control treatments and higher than in the chitosan treatment. Between these days, a constant weight loss was observed in the control treatments and, to a lesser extent, in the other treatments (**Figure 3**). This suggests a regulation of metabolic processes where respiration plays an important role in the loss of water that is related to the weight loss of the fruits [47] and the changes in firmness of the pericarps. The above was reduced after day 6, as was the loss of firmness that was directly proportional to a reduced weight loss in the propolis treatments (regardless of the type of propolis). This suggests that the EEP contains molecules that penetrate the cuticle of the pericarp of the fruit and regulate all these metabolic processes.

The correlation found between the dynamics of phenolic compounds and the dynamics of firmness, being values of 0.40, 0.45, and 0.48 for the respective concentrations of *M. beecheii* propolis and 0.44, 0.48, and 0.50 for the respective concentrations of *S. mexicana* propolis compared with the value of −0.68 of the chitosan treatment and 0.43 and 0.45 of the controls without propolis, suggest that compounds constituents of the EEP were involved in the regulation of the metabolic processes involved in the regulation of firmness and in the generation of phenolic compounds. The above is contrary to the relationship of phenols with weight loss (−0.48, −0.52, and − 0.54 for *M. beecheii*, respectively, of −0.46, −0.48, and − 0.50 for *S. mexicana*, respectively, and for chitosan it was 0.65) and the relationship of firmness with weight loss (−0.53, −0.58, and − 0.62 for the respective concentrations of *M. beecheii* propolis, where for *S. mexicana* they were − 0.48, −0.52, and − 0.56, being for chitosan −0.50), since the values obtained were negative.

These correlation results suggest that one or more components of the EEP, which are possibly antioxidants, penetrated the pericarp of the lemon fruit and acted as regulators of the biochemical and physiological processes involved in (a) regulating the generation of phenolic compounds and maintaining their constant concentration between 29 and 43 mgEAG/g; (b) regulating the changes in the firmness values in the pericarp; and (c) reducing the rate of weight loss at a temperature range of 28–36 degrees Celsius after the sixth day of storage.

Our results differ from those observed by Martínez-González [43] and Báez-Sañudo and Mercado-Ruíz [47], who agree that the greater the amount of phenolic compounds, the greater the weight loss and the drastically reduced firmness values.

5. Conclusions

The application of ethanolic extracts of propolis from *Scaptotrigona Mexicana* and *Melipona becheeii* from Soconusco Chiapas reduces the damage to the leaves of lemon plants (*Citrus aurantifolia*) caused by *Colletotrichum spp*. The damage caused to the leaves of lemon trees was different for each part of the leaf, with the area attached to the petiole being the least damaged. Biofilms based on ethanolic extracts of native bee propolis (*Melipona becheeii* and *Scaptotrigona mexicana*) reduce damage to lemon fruits. The concentration of 15% of EEP from native bees (*Melipona becheeii* and *Scaptotrigona mexicana*) reduces the incidence of diseased fruits by 20%. EEP maintains phenolic compounds within a range of 28 and 43 mgEAG/g, gradually increases the concentration of flavonoids, reduces weight loss, and prevents loss of firmness in lemon fruits.

Acknowledgements

The authors of this chapter would like to thank the owner of the property, accountant Judith Wong Cigarroa, for her support in the form of equipment, land, some supplies, and biological material.

Conflict of interest

The authors declare no conflict of interest.

Author details

José Alfonso López García[1], Julieta Grajales Conesa[1], Víctor Jesús Albores Flores[1*], Maria Guadalupe De Gyves Córdova[1], Rodolfo Torres de los Santos[2], Hermilo Lucio Castillo[2], Marycarmen Utrilla Vázquez[3], Betsabé Rodríguez Pérez[4] and Leopoldo Caridad Cruz López[5]

1 Institute of Biosciences, Autonomous University of Chiapas, Tapachula, Chiapas, Mexico

2 Autonomous University of Tamaulipas, Mante-Center Multidisciplinary Unit, Cd. Mante, Tamaulipas, Mexico

3 Polytechnic University of Tapachula, Tapachula, Chiapas, Mexico

4 National Autonomous University of Mexico, FES Cuautitlan Iztacala, Mexico

5 The College of the Southern Border, Tapachula Unit, Tapachula, Chiapas, Mexico

*Address all correspondence to: alboresflores@gmail.com

References

[1] SIAP. Producción Agricola del gobierno de México. 2021

[2] Sáenz A, Osorio E, Benigno E, Poot A, Delgado R, Rodriguez R. Principales enfermedades en cítricos. Revista Mexicana de Ciencias Agrícolas. 2019;**10**(7):1653-1665

[3] Vincent A, Tuset J. Hongos de cuarentena en frutos cítricos. Cítricos Transferencia y Tecnología. 2011;**230**:38-40

[4] Borja-Bravo M, Velez-Izquierdo A, Cuevas-Reyes V. Rentabilidad y competitividad del limón mexicano en un ambiente endémico de Huanglongbing bajo dos manejos tecnologicos. Ciencias Sociales. 2021;**16**:102-115. DOI: 10.7440/res64.2018.03

[5] Megchún-García JV, Vásquez-Ramírez A, Orozco-Corona DM, Milagro-Morales AG, Robles I, Espinoza J. Comercialización de Limón (Citrus limón) en la zona sur del estado de Veracruz. Ciudad de México: SADER, INIFAP, Centro de Investigaciones Nacionales del Golfo; 2022. 500 p. ISBN: 978-607-37-1360-3

[6] Ordoñez-Gómez ES, Reátegui-Díaz D, Villanueva-Tiburcio JE. Polifenoles totales y capacidad antioxidante en cáscara y hojas de doce cítricos. Scientia Agropecuaria. 2018;**9**(1):113-121. DOI: 10.17268/sci.agropecu.2018.01.13

[7] Dong X, Hu Y, Li Y, Zhou Z. The maturity degree, phenolic compounds and antioxidant activity of Eureka lemon [Citrus Limon (L.) Burm. f.]: A negative correlation between total phenolic content, antioxidant capacity and soluble solid content. Scientia Horticulturae. 2019;**243**:281-289. DOI: 10.1016/j.scienta.2018.08.036

[8] Panahirad S, Dadpour M, Peighambardoust SH, Soltanzadeh M, Gullón B, Alirezalu K, et al. Applications of carboxymethyl cellulose- and pectin-based active edible coatings in preservation of fruits and vegetables: A review. Trends in Food Science and Technology. 2021;**110**:663-673. DOI: 10.1016/j.tifs.2021.02.025

[9] Herrera-López MG. Caracterización química y actividad biológica de propóleos producidos en el estado de yucatán [In Tesis de Doctorado]. Editorial: Centro de Investigación Cientifica de Yucatán A.C. Posgrado en Ciencias Biológicas; 2019;**201**

[10] Rebaza R, Amaya L, Gutiérrez A, Haro R, Tumbajulca M, Valera F, et al. Aplicación del propóleo en envasado activo. Agroindustrial Science. 2017;**6**(2):239-252

[11] Viloria-B JD, Gil-G JH, Durango-R DL, García-P CM. Caracterización Fisicoquímica Del Propóleo De La Región Del Bajo Cauca Antioqueño (Antioquia, Colombia). Biotecnología En El Sector Agropecuario y Agroindustrial. 2012;**10**:77-86. Available from: http://www.scielo.org.co/pdf/bsaa/v10n1/v10n1a10.pdf

[12] Espejel-Gaspar X, Hernández-López G, Chirino-Galindo G, Palomar-Morales M. Elaboración y evaluación de una biopelícula comestible con el uso de grenetina, miel y cera de abeja, para la conservación de tres variedades de Vitis vinifera (uva). Universidad Autónoma Metropolitana Revista TeDIQ. 2020;**6**(6):592-598

[13] Fernández Valdés D, Bautista Baños S, Fernández Valdés D, Ocampo Ramírez A, García Pereira A, Falcón

Rodríguez A. Películas y recubrimientos comestibles: una alternativa favorable en la conservación poscosecha de frutas y hortalizas. Revista Ciencias Técnicas Agropecuarias. 2015;**24**(3):52-57

[14] Quezada MD, Segovia GJ, Añazco MV. Desarrollo de película biodegradable a base de propóleo para la conservación de mango (Tommy Atkins). Alimentos Hoy. 2018;**26**(44):30-43

[15] Devi LS, Jaiswal AK, Jaiswal S. Lipid incorporated biopolymer based edible films and coatings in food packaging: A review. Current Research in Food Science. 2024;**8**:100720. DOI: 10.1016/j.crfs.2024.100720

[16] Muñoz-Tebar N, Pérez-Álvarez JA, Fernández-López J, Viuda-Martos M. Chitosan edible films and coatings with added bioactive compounds: Antibacterial and antioxidant properties and their application to food products: A review. Polymers (Basel). 2023;**15**(2):396. DOI: 10.3390/polym15020396

[17] Lorini A, Wobeto C, Bonaldo SM, de Carvalho Campos S, Sinhorin AP. Chemical composition and antifungal activity of propolis on Aspergillus flavus. Bioscience Journal. 2018;**34**(5):1298-1307. DOI: 10.14393/BJ-v34n5a2018-36601

[18] Chakraborty B, Shashiraj KN, Basavarajappa DS, Bhat MP, Nayaka S, Kumar RS, et al. Biological activities and GC–MS based chemical profiling of polymolecular methanol extract of Alternaria alternata KUDB15. Journal of King Saud University - Science. 2023;**35**:8. DOI: 10.1016/j.jksus.2023.102884

[19] Putra RE, Reizandy F, Faizal A, Kinasih I. Effication of local propolis as edible coating of tangerine cultivar Garut (Citrus reticulata Blanco). IOP Conference Series: Earth and Environmental Science. 2018;**187**(1). DOI: 10.1088/1755-1315/187/1/012025

[20] Poverenov E, Danino S, Horev B, Granit R. Layer-by-layer electrostatic deposition of edible coating on fresh cut melon Model : Anticipated and unexpected effects of alginate – Chitosan combination. Food and Bioprocess Technology. 2013;**7**:1424-1432. DOI: 10.1007/s11947-013-1134-4

[21] Barrera-Bello E, Loaiza MG, Mario C, Pajón G, Luis D, Restrepo D, et al. Empleo de un Recubrimiento Formulado con Propóleos para el Manejo Poscosecha de Frutos de Papaya (Carica papaya L. cv. Hawaiana) use of a coating formulated with propolis for postharvest handling of papaya (Carica papaya L. cv. Hawaiian) fruits. Revista Facultad Nacional de Agronomía. 2012;**65**(1):6497-6506

[22] Osís-Rojas AG. Efecto del Recubrimiento de Propóleo Natural de Abeja en la Pérdida de Peso, Color, Firmeza y Ánalisis microbiooligco en la Pal Hass (Persea Americana Miller) [In Tesis de Licenciatura]. Editorial: Universidad Nacional Jóse María Arguedas; 2020;**202**

[23] Domínguez-López RM. Hipersensibilidad a Biopelículas de Propóleo para conservación de Frutas [Thesis de grado]. Universidad de Valladolid, Facultad de Medicina; 2022. Available from: https://uvadoc.uva.es/handle/10324/54971

[24] Rico-Lopez E. Optimización de un método de obtención de extracto etanólico de propóleo en base al poder antimicrobiano [Tesis de Maestria]. Editorial: La Universidad Politécnica de Valencia; 2017

[25] Pérez-Torres BC, Aragón-García A, Cuate-Mozo VA, López-Olguín JF, Aragón-Sánchez M, Lugo-García G.

Efecto de la aplicación en campo de mezclas de extractos vegetales sobre la presencia y daños de insectos plaga del cultivo de Amaranthus hypochondriacus L. Revista de la Facultad de Agronomía. 2017;**34**(4):477-496

[26] Oroz-Santos M, García-Mariscal K, Ázquez-Jiménez JL, RoblesGonzález MM. Control de plagas y enfermedades con aspersión de aceite refinado en el cultivo de limón Mexicano (Citrus aurantifolia) Mario. In: Paper Knowledge. Toward a Media History of Documents. Editorial: Instituto Nacional de Investigaciones Forestales, Agrícolas y Pecuarias. 2008;**5**(2):40-51

[27] Virgen-Hernández CA, Albores-Flores VJ, Grajales-Conesa J, Vázquez-Ovando JA. Efecto de biopelículas de quitosán adicionadas con cera y miel de abejas sin aguijón en frutos de mango sobre la incidencia de Colletotrichum gloeosporioides [In Tesis de Licenciatura]. Editorial: Universidad Autónoma de Chipas, Instituto de Biociencias; 2021

[28] del Carmen Martínez-González M, Bautista-Baños S, Correa-Pacheco ZN, Corona-Ragel ML, Ventura-Aguilar RI, Río-García JC, et al. Effect of nanostructured chitosan/propolis coatings strawberries during storage. Coatings. 2020;**10**:1-12

[29] Peláez-Sánchez PP, Roldan-Carbajal W, Carmona-Ruiz A, Raza-Quiroz RC. Evaluación de la Capacidad Antioxidante y Estabilidad Térmica de la Cáscara, Zumo y Semilla de la Lima Dulce, Limón Rugoso y Limón Tipo Mandarina. Investigación y Amazonía, Tingo María, Perú. 2019;**9**(7):38-46

[30] Amaya L, Portillo C. Determinacion de fenoles, flavonoides y capacidad antioxidante en melaza, azucar blanco y moreno en el Ingenio Chaparrastique por el método de espectrofotometria Ultravioleta-Visible [Tesis de Pregrado]2013. pp. 53-54. Available from: https://bit.ly/2msX9o5

[31] Cavalho GJL, Sodré GS. Application of propolis in agriculture. Arquivos do intituto biológico. 2021;**88**(1-8):e0632019. DOI: 10.1590/1808-1657000632019

[32] Cibanal I, Fernández L, Krepper G, Pellegrini C, Gallez L. Avances en el desarrollo de un biofungicida: caracterización físico-química y actividad antifúngica de propóleos. Agrociencia (Uruguay). 2019;**23**(2):99-108. DOI: 10.31285/agro.23.83

[33] Retamoso RM, Ruíz GB, Benítez RM. Efecto antifúngico de extractos de propóleos obtenidos en la provincia de Jujuy, Argentina. Revista Científica de la Facultad de Ciencias Agrarias. 2021;**14**(1):69-76

[34] Joya M, Gil M, Bastidas-Pacheco G. Actividad fungistática y fungicida de extractos etanólicos de propóleos sobre el crecimiento in vitro de cepas del género Candida. Tecnología en Marcha. 2017;**30**(3):3-11

[35] Romero Rivera MH. Interacciones entre compuestos mayoritarios anti-Helicobacter pylori presentes en propóleos de la Región del BioBío [Tesis Doctoral]. Chile: Universidad de Concepción; 2017

[36] Rodríguez Pérez B, Canales Martínez MM, Penieres Carrillo JG, Cruz Sánchez TA. Composición química, propiedades antioxidantes y actividad antimicrobiana de propóleos mexicanos. Acta Universitaria. 2020;**30**:e2435. DOI: 10.15174/au.2020.2435

[37] Tafolla-Arellano JC, González-León A, Tiznado-Hernández ME,

Zacarías García L, Báez-Sañudo R. Composición, fisiología y biosíntesis de la cutícula en plantas. Revista Fitotecnia Mexicana. 2013;**36**(1):3-12. Available from: http://www.scielo.org.mx/scielo.php?script=sci_arttext&pid=S0187-73802013000100001&lng=es&tlng=es

[38] Pratyusha S. Phenolic compounds in the plant development and defense: An overview. 2022. DOI: 10.5772/intechopen.102873

[39] Ghaemi Z, Alberga D, Carloni P, Laio A, Lattanzi G. Permeability coefficients of lipophilic compounds estimated by computer simulations. Journal of Chemical Theory and Computation. 2016;**12**(8):4093-4099. DOI: 10.1021/acs.jctc.5b01126

[40] Brizzolara S, Manganaris GA, Fotopoulos V, Watkins CB, Tonatti P. Primaru metabolism in fresh fruit during storage. Frontiers in Plant Science. 2020;**11**:80. DOI: 10.3389/fpls.2020.00080

[41] Amodio ML, Derossi A, Colelli G. Modeling phenolic content during storage of cut fruit and vegetables: A consecutive reaction mechanism. Journal of Food Engineering. 2014;**140**:1-8. ISSN 0260-8774

[42] Oliveira A, Almeida D, Pintado M. Changes in phenolic compounds during storage of pasteurized strawberry. Food and Bioprocess Technology. 2014;7. DOI: 10.1007/s11947-013-1239-9

[43] Martinez-González ME, Balois-Morales R, Alia-Tejacal I, Cortes-Cruz MA, Palomino-Hermosillo YA, Lopez-Gúzman GG. Poscosecha de frutos: maduración y cambios bioquimicos. Revista Mexicana de Ciencias Agricolas. 2017;**8**:4075-4087. DOI: 10.29312/remexca.v0i19.674

[44] Anaya-Esparza LM, Pérez-Larios A, Ruvalcaba-Gómez JM, Sánchez-Burgos JA, Romero-Toledo R, Montalvo-González E. Funcionalización de los recubrimientos a base de quitosano para la conservación poscosecha de frutas y hortalizas. Revista Especializada en Ciencias Químico-Biológicas. 2020;**23**:1-14. DOI: 10.22201/fesz.23958723e.2020.0.241

[45] Xoca-Orozco LA, Cuellar-Torres EA, González-Morales S, Gutiérrez-Martínez P, López-García U, Herrera-Estraella L, et al. Transcriptomic analysis of avocado Hass (Persea americana Mill) in the interaction system fruit-chitosan-colletotrichum. Frontiers in Plant Science. 2017;**8**:956. DOI: 10.3389/fpls.2017.00956

[46] Mulkay-Vitón T, Paumier-Jiménez A, González-Vasallo J. Efecto de extractos de propóleos en la calidad poscosecha del mango Super Haden. Fitosanidad. 2016;**20**:5-11

[47] Baéz-Sañudo R, Mercado-Ruíz JN. Estudios para el control de desarrollo de cavidades en la pulpa del mango "ataulfo" en el suroeste de México. Revista Iberoamericana de Tecnología Poscosecha. 2023;**24**:1-15

Section 3

Agricultural Systems and Sustainability

Chapter 7

Organic Fertilizer: Indonesia's Legacy for a Sustainable Future

Ratih Kemala Dewi, Restu Puji Mumpuni,
Muhammad Iqbal Nurulhaq, Agief Julio Pratama,
Edi Wiraguna, Ririh Sekar Mardisiwi, Widya Hasian Situmeang,
Tri Budiarto and Henry Kasman Hadi Saputra

Abstract

In the current state of soil degradation, organic fertilizers are crucial for sustainable agricultural production in Indonesia. As Indonesia is a major agricultural production, this review that mainly comes from locally published papers from Indonesia reported that organic fertilizers are crucial for sustainable agricultural production in Indonesia. They have been used in natural farming practices long before Indonesia was established. Organic fertilizers come in various forms, including compost (crop residues, manure, vermicompost, and maggot compost), green manure, biofertilizers, and biochar. They are primarily used as soil amendments to enhance soil and plant health. Organic fertilizers can be combined with inorganic fertilizers or applied alone. For effective organic farming, organic fertilizers must be managed in conjunction with other practices to boost crop productivity and economic value. The long tradition of organic farming and fertilizers is upheld by local communities. The Indonesian government continues to provide assistance and support to promote the use of organic fertilizers, enhancing agricultural sustainability.

Keywords: crop quality, indigenous knowledge, local communities, soil amendment, soil health, sustainability

1. Introduction

Indonesia, a major agricultural nation located in the tropics, is a significant global food producer. Agriculture stands as a crucial sector of the Indonesian economy. The agricultural sector contributed 12.4% to the national gross domestic product (GDP) dominated by plantation commodities, food crops, horticulture, livestock, and fisheries [1]. Even though this GDP has decreased, the agriculture sector remains a vital income generator for the majority of Indonesian households. Around 40.7 million Indonesians, or 29.4% of the total labor force, were employed in agriculture [1]. Given its essential role, best management practices involving organic amendments are key to ensuring the sustainability of Indonesian agriculture.

Organic materials, including crop residue compost and manure, are the most common sources used in agricultural land. Traditionally, crop residues have been returned to the land for centuries. However, as intensification in agriculture production increased, the use of organic materials began to decline. Particularly during the Green Revolution, most farmers preferred synthetic fertilizers to boost crop yields, leading to issues such as land degradation and farmers' dependence on inorganic fertilizers. However, with the growing awareness of sustainable agriculture and healthier living, the demand for organic products has increased. The Indonesian government has focused on improving degraded land and encouraging the use of organic fertilizers to enhance soil health and support plant growth. Many studies have explored alternatives to organic fertilizers that are efficient and easy to apply, such as green manure, biofertilizers, and biochar, as well as methods to produce compost with better nutritional quality, such as vermicomposting.

This study explores organic materials used for fertilizer and their impact on soil health—encompassing physical, chemical, and biological properties—and crop yield in Indonesia. It also delves into the history and local wisdom of organic farming based on the Indonesian experience and tradition. Moreover, this study also provides information on the production and distribution of organic fertilizer in Indonesia, highlighting governmental efforts to promote and support the adoption of new technologies by farmers. Overall, this review aims to provide comprehensive insights into the role of organic fertilizers in advancing sustainable agriculture in Indonesia and globally.

2. History of organic fertilizer in Indonesia

Long before the Indonesian nation was formed, organic farming system was practiced by local residents as stated in the ancient manuscript below in **Table 1**:

The meaning of the two quotes above is "on the edge of the village; it's good to be shady; lots of coconuts and guavas; durian trees jackfruit trees; bamboo behind the house; don't forget to plant aren palm; because it makes the shade; and who plant a lot of coffee; under the trees; there is no empty place; and it is necessary to make a dug-up place; for garbage disposal; collected into fertilizer; if the garbage has rotted", then in the next verse "dry trees; must fertilize the soil; and again according to the

215.11	216.12
Padhedhusunan ing pinggir	*Uwit-uwitan kang gering*
Abecik kang iyup padha	*Kudu den rabuk lemahnya*
Kebak kelapa lan jambu	*Lan maneh dhokter trekahe*
Uwit duren uwit nangka	*Ing desa kang resik padha*
pring ing burining omah	*Amerga dadi tulak*
Nandur aren saja suwung	*Lara koreng gudhik iku*
Marga becik ing keyuban	*Kolerah uga tan prapta*
Sarta den keh nandur kopi	Verses 215 dan 216 "*Wulang Tani*" by Raden Haji Moehamad Moesa (1860s) [2]
Ing ngisoring uwit-witan	
aja ana papan lowong	
Lan prelu gawea blumbang	
Kinarya gon pawuhan	
Kinumpukken karya rabuk	
Yen wus abosok uwuhnya	

Table 1.
The part of an ancient manuscript that represent natural farming in Indonesia.

doctor; in a clean village; because it becomes a repellent; sick '*koreng gudhik*' (like scabs); cholera will also not reach". [2]. These are two pieces of verse from the book "*wulang tani*" where "*wulang*" means instructions and "*tani*" means farming. So wulang tani means "instructions for farming". This is a part of an ancient manuscript by a poet in Sundanese land who was a village head (*Hoofdpangulu*) named Raden Haji Moehamad Moesa (1822–1886).

In both verses, it mentions the instructions for making compost by making a digging place for garbage disposal (leaves), which will become fertilizer (compost) when it decomposes. The next verse talks about the importance of fertilizing the soil for dry trees. This ancient manuscript is a guide to organic farming that has become a reference in agricultural learning in the Sundanese region of Java and Madura, Indonesia.

Agricultural practices using organic fertilizers began to be gradually replaced since the discovery of ammonia synthesis process technology by Fritz Haber and Carl Bosch in 1909, known as the Haber-Bosch Process. The discovery made it possible to produce synthetic nitrogen fertilizer cheaply. On its way, the production of synthetic nitrogen is also used as an explosive for the needs of World War I [3].

In 1924, an Austrian philosopher named Rudolf Steiner (1861–1925) made a breakthrough to return agriculture to a more natural agriculture through organic farming. This was a response to the rapid practice of "chemical" agriculture. That year saw the start of the first training for organic farming, although the term organic farming only emerged decades later [4]. The first lessons in the training by Steiner were prioritized on procedures in the preparation of biodynamic fertilizers. Biodynamics itself is a spiritual approach that involves cosmological elements with the ideological roots of Anthroposophy, which consists of the words "Anthropos" (human) and "sophia" (wisdom), so when combined, it means Human Wisdom [5]. Through this approach, agriculture is considered a living organism.

Biodynamic agriculture includes environmentally friendly organic farming practices through specific farming systems with biodynamic preparations that are believed to be able to connect cosmic energy with agriculture [5]. One form of the practice of using organic fertilizers in biodynamic agriculture is the preparation for improving soil fertility. The preparation uses cow dung, leaves, flowers, silica, and other organic materials that are mixed with different techniques and applications (soil or plants) [6].

3. Organic fertilizer and its impact on soil, plant, and environment

3.1 Compost and animal manure

The addition of compost is very common among farmers as a source of organic matter. Organic fertilizers in a composted form play an important role in improving the chemical, physical, and biological properties of soil and serve as a source of plant nutrients. Sources of compost materials include organic waste such as crop residues (straw, stems, branches), household waste, livestock manure (cows, goats, chickens), husk charcoal, kitchen ash, and forage. The results of the study by Mahbub et al. [7] showed that organic fertilizer significantly increased the available potassium (K) content of paddy soil, but was not able to significantly increase the pH, total nitrogen (N), organic carbon (C), and available phosphate (P) of the soil. Integrated application of organic fertilizers with reduced doses of inorganic fertilizers can spur rice growth and yield [8]. The use of straw compost can reduce the use of inorganic fertilizers [9]. The combination of 50% inorganic and 50% organic fertilizers can increase the yield of rice by 23.8% compared to without adding organic fertilizers. Organic fertilizers can improve plant root development, so it is expected to increase the efficiency of inorganic fertilizers.

Research by Razavipour et al. [10] showed that giving Azolla compost at a rate of 5.0% of soil weight resulted in higher yields, with an average increase of 13.8% compared to the control without Azolla application. This rate also produced a higher number of tillers, panicle weight, and grain yield compared to the control without Azolla application. Therefore, a rate of Azolla of 5.0% by weight of soil can be considered the best rate in this study. The use of vermicompost has been proven to have a positive influence on rice growth and yield. Hoque et al. [11] showed that the combination of mineral fertilization with vermicompost can improve characteristics that contribute to rice yield, including grain and straw yield (**Table 2**). The study also showed that the use of vermicompost can increase the availability of soil nutrients and the absorption of nutrients by rice plants. Application of 10 tons ha^{-1} of vermicompost on rice plants can increase grain yield by 25% compared to the control treatment.

Manure is a type of organic fertilizer that is a mixture of solid manure, urine, and animal (plant) food scraps. Thus, its chemical composition depends on the type of livestock, the age and condition of the animal, nature, and the way the fertilizer is stored before use. Some of the solids contained in manure consist of organic compounds similar to food ingredients, including cellulose, starch, sugars, hemicellulose, and lignin as we find in ligno-protein humus. In general, the nutrient content in animal manure is lower than that of inorganic fertilizers. Therefore, the dosage of manure is much larger than that of inorganic fertilizers. Apart from being a source of nutrients, organic fertilizer also functions as a soil improver. In addition to containing nutrients needed by plants, manure also contains humic acids, fulvic acids, growth hormones, and others that spur plant growth so that nutrient uptake by plants increases. Organic fertilizers, such as fertilizers from poultry manure containing high total N (>4%) and moderate P and K, have more influence on crop yields [8]. Poultry manure has more influence on several rice growth parameters than the use of cow manure [8]. In their research, they used a dose of 5 tons ha^{-1} of cow manure, and the dose of poultry manure used was 5 tons ha^{-1}. Some rice growth parameters that can be affected by the use of poultry manure fertilizer include 1. Plant height: Poultry manure fertilizer can affect the vertical growth of rice plants, which is reflected in better plant height. 2. Number of tillers: The use of poultry manure fertilizer can also affect the number of tillers in rice plants, which is an important indicator of plant productivity. 3. Chlorophyll content: Poultry manure fertilizer can affect chlorophyll content in plants, which is an indicator of photosynthesis rate and plant health. Total N, P, K, and calcium (Ca) in poultry manure were higher than in cow manure and compost (**Table 3**).

Organic manures	Moisture (%)	Organic C (%)	N (%)	P (%)	K (%)	S (%)
Vermicompost (n = 8)	68.0 ± 3.1	34.0 ± 2.1	1.68 ± 0.16	0.4 ± 0.06	1.3 ± 0.12	0.5 ± 0.06
Cow dung (n = 3)	78.3 ± 3.5	45 ± 2.6	1.63 ± 0.19	0.4 ± 0.09	0.9 ± 0.15	0.6 ± 0.07

Table 2.
Comparison of nutrient content in vermicompost and cow dung.

Organic manures	Ash	Moisture	Total N	NH_4^+–N	NO_3^-–N	P_2O_5	K_2O	Ca	Mg	Na
Poultry manure	37.4	21.9	4.9	0.8	0.0	4.6	2.1	8.9	0.8	0.3
Cow manure	31.5	46.4	2.4	0.2	0.0	1.9	1.5	1.1	0.6	0.3
Compost	28.1	41.0	2.2	0.04	0.02	2.3	1.2	5.3	0.2	0.9

Table 3.
Chemical compositions of poultry manure, cow manure, and compost (%).

Bachtiar et al. [12] showed that the application of manure can increase the contribution of nutrients, especially N from the soil in paddy rice. The application of 15 tons ha^{-1} of manure plus N recommendations had a significant effect in increasing the contribution of N from the soil in straw by 19.98% from the treatment without manure plus N. The results showed that the higher the application of animal manure, the higher the yield and uptake of N. Wu et al. [13] reported that the application of chemical fertilizers together with the application of manure is an effective method in increasing available soil N. This occurs due to increased mineralization of N in the soil. However, N nitrification can also be higher due to the induction of manure, so it can also cause an increase in N loss. Rahimabadi [14] revealed that the application of cow dung and vermicompost can increase chlorophyll in leaves and yield components such as the number of tillers and the number of full grains.

3.2 Vermicompost

Vermicomposting, a technique using worms to convert organic waste into nutrient-rich fertilizer, has roots in ancient practices, with Mesopotamians first utilizing worms to enhance soil fertility. However, it gained global recognition in the late twentieth century, especially with the advent of modern vermicomposting. This was advanced by Fletcher Sims Jr.'s windrow method and studies on efficient earthworm species like *Eisenia fetida* (red worms) and *Lumbricus rubellus*. [15]. Vermicomposting was introduced in Indonesia in the 1990s as awareness grew about its environmental benefits and the need for organic waste management. Institutions like Bogor Agricultural University (IPB) began researching its impact on soil fertility and agricultural productivity. Supported by government initiatives promoting organic waste management, more farmers adopted organic farming, using vermicomposting to convert agricultural waste into high-quality organic fertilizer, enhancing its role in addressing waste issues, and boosting agricultural yields.

3.2.1 Decomposition interval of organic substances into Vermicompost

Beyond the breakdown of organic materials, vermicomposting entails crucial physical and chemical changes to create high-quality compost. Analyzing changes in the C/N ratio and volatile solids over time is a crucial step that provides information on process effectiveness and affecting variables. This knowledge promotes sustainability in agriculture and natural resource management while improving waste management techniques.

According to Norbu [16], the vermicomposting process over 2 weeks showed consistent reductions in both volatile solids and the C/N ratio. Volatile solids decreased from 68% on the first day to 55% by day 16, with an overall reduction of 19.11%. The C/N ratio dropped from 21.46 on day one to 16.16 by day 16, marking a 24.69% total decrease. These findings highlight vermicomposting's efficiency in reducing organic waste and improving waste management practices. Comparing vermicompost to conventional composting, worms are crucial in drastically lowering the C/N ratio, suggesting higher carbon reduction efficiency. Higher amounts of exchangeable nutrients are also present in vermicompost, which increases its potency as a fertilizer for plant growth. Although it needs more careful handling, vermicompost provides better nutrient availability and more advantages for plants.

3.2.2 Mechanism of decomposition of volatile solid waste in vermicompost

Vermicompost relies on worms as key agents in decomposing volatile solid waste into nutrient-rich compost. To produce high-quality vermicompost that is beneficial for plants [17], two main factors must be managed: (1) pH levels: Earthworms thrive at a neutral to slightly acidic pH (6.5–7.5), making pH regulation essential for their survival and optimal activity; (2) temperature: The ideal range is 20–30°C, as unsuitable temperatures can reduce productivity by up to 30% or cause worm mortality. Earthworms reduce the C/N ratio through several mechanisms: (1) physically crushing organic particles and secreting mucus, increasing the surface area for digestion; (2) enzymatic and microbial activity in their digestive tract breaks down complex carbon compounds into simpler forms, converting carbon into volatile forms like CO_2; (3) as CO_2 is released, carbon content decreases while nitrogen is retained in the earthworm's output. The role of worms is crucial in enabling vermicompost to achieve a significant decrease in the C/N ratio when compared to standard composting, suggesting that the vermicomposting process is more effective in reducing the carbon content found in regular compost materials. Furthermore, vermicompost possesses advantages in terms of nutrient content that can be exchanged or readily absorbed by plants.

3.2.3 Benefits of vermicompost in comparison to traditional compost

Vermicompost offers significant advantages over conventional compost, benefiting plants more effectively. According to Norbu [16], vermicompost: (1) reduces volatile solids by 55–68%, about 10–15% more than aerobic composting, making it more efficient in stabilizing organic material; (2) decreases the C/N ratio by 20–30%, compared to 10–20% with aerobic composting, indicating better carbon breakdown and conversion to plant-usable forms; and (3) increases exchangeable nutrients, with potassium and phosphorus levels 20–25% higher than conventional compost (10–15%). Additionally, vermicompost has balanced properties beneficial to plants, including a pH of 6.8–7, organic carbon content of 25.4–27.5%, nitrogen levels of 1.2–1.6%, and other advantageous parameters.

Vermicomposting effectively enhances available phosphorus (P) from organic waste [18] and outperforms conventional composting in nutrient availability, producing higher-quality outputs [19]. This odorless, economical, and eco-friendly method yields manure with more accessible nitrogen. Hammermeister et al. [20] found vermicompost richer in nitrates, a plant-available nitrogen source, compared to the higher ammonium content in conventional compost [21]. Vermicompost is a tried-and-true technique for boosting nutrients from organic materials, resulting in fertilizers of higher quality than conventional fertilizers. The fertilizer made from earthworms has a free odor, is reasonably priced, and increases nitrogen availability by making nitrate levels higher and easier for plants to absorb.

3.3 Maggot BSF compost (Maggcompost)

The technology of fertilizer, particularly compost made from the bioconversion of organic waste, has evolved significantly during the last 10 years. Maggcompost, also known as Bekas Magot in Indonesian, is made from the Black Soldier Fly (BSF) maggots (*Hermetia illucens*). The benefits of maggcompost are similar to those of other compost varieties, but its special benefits are still mostly overlooked. The BSF life cycle has six stages—egg, mini larva, larva, prepupa, pupa, and adult—lasting about 44 days. Eggs hatch 3–4 days after the female lays over 400 eggs. Mini larvae (<1 cm), while larger larvae (>1 cm, brownish), efficiently process waste over 7–14 days. Prepupae (14–29 days) do not process waste and seek dry areas. During the pupal phase (30–44 days), larvae stop eating and become adult flies. The cycle's speed depends on food quality, with finer, nutritious food accelerating growth. BSF larvae convert organic waste in 1 day, faster than traditional composting (24–45 days). One kg of larvae processes 5–10 tons of waste, yielding 0.1 tons of larvae and 0.2–0.3 tons of maggcompost per ton of waste. This method offers a quick, efficient, and sustainable waste management and fertilizer solution [22]. Maggcompost supports plant development, metabolism, and stress resistance by supplying vital macronutrients (N, P, K) and micronutrients (Ca, Zn, Fe, Cu, Mg, B). Additionally, it has natural bio-stimulants that improve plant resistance and nutrient uptake.

Bio-stimulants are natural substances that enhance plant growth, health, and productivity, often outperforming conventional fertilizers. Found in maggcompost, bio-stimulants include chitin, a chitin-microorganism combo, and microorganisms. [23]. Maggcompost with chitin at 7.5 and 15 g kg^{-1} effectively reduces *Pythium ultimum* and suppresses Fusarium wilt due to chitin's antimicrobial properties, which protect plant roots and inhibit soil pathogens. Amino acids, humic acid, and fulvic acid also improve soil structure and nutrient absorption. Maggcompost provides an eco-friendly approach to plant health, reducing the need for chemical pesticides [24]. Its chitin and microorganism combination enhances soil quality, enriches nutrients, and suppresses pathogens like *Rhizoctonia*, *Fusarium*, and *Pythium*. These microorganisms also strengthen plant disease resistance and increase crop yields [25].

Maggcompost combines chitin and nitrogen-fixing bacteria to boost plant growth, accelerate flowering, and support sustainable farming. It enhances soil fertility, reduces the need for chemical fertilizers, and increases productivity naturally [26]. Maggcompost is an eco-friendly solution that supports sustainable agriculture by enhancing soil fertility, suppressing pathogens, and strengthening plant resistance. It contains chitin, microorganisms, and bio-stimulants, enriching nutrients and reducing reliance on chemical fertilizers.

3.4 Green manure

Green manure is an agricultural practice where specific crops, typically legumes or green leafy plants, are grown and then incorporated back into the soil to enhance organic matter content, soil structure, and microbial activity [27]. Green manure is quite common in Indonesia, with studies on its potential and application dating back a long time. It is widely recognized as a nutrient source for plants by increasing the organic matter in the soil. However, its use has been limited due to the Green

Revolution's focus on food production intensification. Additionally, since green manure production does not generate direct income, many farmers are reluctant to adopt this technology. Integrating grain legumes in the existing cropping system without disturbing the main crops in the system appears to have a greater chance of adoption and may also provide long-term economic benefits to low-land rice-based systems. The economic feasibility of green manure can be improved by enhancing its use as food and feed, which will improve its short-term advantage [28]. Despite its limited use, green manure offers environmental benefits through ecological improvement. In line with sustainable agricultural production, the Indonesian government continues to promote its potential to improve soil health. Green manure can be applied to food crops, horticulture, and plantation production. The utilization of green manure could reduce the chemical fertilizer and improve crop production.

Many shrubs' leaves that come from Leguminosae can be used as nutrient source, cover crop under plantation trees (rubber, coffee, cocoa, oil palm, coconut, citrus, sisal, clove, cinchona), or as soil improvement, for example, *Acacia villosa* Willd., *Arachis pintoi*, *Calliandra calothyrsus* Meissn., *Calopogonium caeruleum* (Benth), *Sauvalle*, *Clitoria ternatea*, *Mucuna bracteata*, *Pueraria phaseoloides*, *Sesbania sesban* (L.) Merrill, *Tephrosia candida* (Roxb.) DC., *Tephrosia vogelii* Hook [29–31]. Additionally, Leguminosae trees, such as *Albizia falcataria* (L.) Fosberg, *Gliricidia sepium* (Jacq.) Walp, *Leucaena leucocephala* (Lam.) de Wit, *Sesbania grandiflora* (L.) Poir, can be used as shade for plantation trees [30].

Most green manure plants are from the legume family due to their ability to fix N_2 from the atmosphere. However, ferns and grasses can also be used as green manure. *Azolla pinnata* (Salviniaceae) can be used to add organic matter to rice fields [30]. Azolla also can be used as a substitute for animal manure in organic farming. A study reported that the use of Azolla 50% + cow manure 50% in a silty clay soil texture with low C content could increase rice growth and this application was equivalent to 100% cow manure application on organic C, total N, and C/N of soil. Utilization of Azolla as a substitution for cow manure in organic rice farming has advantages for utilizing local potential, as it can be cultivated directly in the field and applied to the fields without composting. Using the combination of Azolla 50% + manure 50% can increase organic C content of soil from 1.3 to 1.7%, leading to an improvement in soil health [31]. Green manure may fit into rice farming systems in two broad ways. For all their diversity in terms of species and farming systems, soil-improving legumes may be defined as either pre-rice or post-rice green manure [32]. International rice research institute (IRRI) evaluated several foods + forage crops intercropped at the performance of cereal crops (maize, *Zea mays,* and sorghum, *Sorghum bicolor*), food legumes (cowpea, *Vigna unguiculata*, mungbean, *Vigna radiata*, and pigeonpea, *Cajanus cajan*), and forage legumes (desmanthus, *Desmanthus virgatus*, sunn hemp, *Crotalaria juncea*, clitoria, *Clitoria ternatea*, stylo, *Stylosanthes guianensis*, and siratro, *Macroptilium atropurpureum*) after rice, and their effects on the succeeding rice crop. The studies demonstrated the use of forage legumes as green manure in the rice system to produce quality feed for animals during the dry season when feed is limited and as the last forage regrowth to serve as green manure for the succeeding rice crop. All the studies indicated that food-forage crop production systems can increase cereal yield, produce legume grain crops and forage legume, especially during the dry season when animal feed is limited, decrease the use of inorganic fertilizer, and increase total fodder available to animals in a small-scale production system [32]. Many shrubs and trees contribute high-N leaf litter as nurse trees or are lopped to provide foliage for green manure, compost addition, or fodder including

Acacia, Albizia, Calliandra, Desmanthus, Leucaena, Prosopis, and Samanea [30]. Those genera also act as soil restoration species [30]. Another study also reported that fertilization of N, P and application of Crotalaria green manure improved growth and increased the yield of maize. Fertilization with a dose of 135 kg N ha^{-1} + 66 kg P_2O_5 ha^{-1} with the application of Crotalaria obtained a yield of maize as many as 7.2 tons ha^{-1} while fertilization at the same dose without the application of Crotalaria obtained the yield as many as 6.2 tons ha^{-1} [33]. The current study also reported that compared to mung bean and peanut residue, *Centrosema pubescens* compost at a rate of 7.5 tons ha^{-1} can improve the growth and yield of maize [34].

Similar to food production, vegetable production requires a high amount of nutrients and even more intensive maintenance. To stabilize the inputs and outputs of the system, organic fertilizer is often needed to improve soil health and provide nutrients to the plants. Green manure can be used as a complementary source of organic fertilizer. In organic farming systems, vegetables are primarily grown using mixed crops and crop rotation. The land is not continuously cultivated but is given time to rest. During the fallow period, legume cover crops (LCC) are planted, and these cover crops are later terminated to serve as green manure for the subsequent crop.

In plantation production, green manure is applied as a cover crop or as a tree shade for coffee and cocoa. Leguminous cover crops can be used to minimize soil erosion and maintain soil fertility in oil palm plantations. Planting LCC not only helps replenish soil organic matter stock but also reduces the velocity of soil and water movement. The LCC is commonly established in oil palm estates including *Mucuna bracteata*, *Pueraria phaseoloides*, and *Calopogonium caeruleum*. In steep terraced areas, deep rooting Vetiver grass can be used to prevent soil erosion and regulate water excess [35]. The growth of oil palm (with LCC) tends to counteract the negative impacts of clearance, without always returning to pre-disturbance levels. Along with the application of palm oil effluent mill (POME) and empty fruit bunch (EFB) mulch, the establishment of LCC helps to reduce chemical fertilizer used. In the rubber plantation, *Mucuna bracteata* could increase soil organic matter content by 30.43% in flat areas and 53.33% in hilly areas, amount of total N by 27.3% in flat areas and 7.7% in hilly areas, soil infiltration by 48.9% with sandy clay dominant soil texture. Moreover, the *Mucuna bracteata* could reduce the soil bulk density [29]. Many Leguminosae can be used as fodder and cover crops such as *Clitoria ternatea*. It is considered a fodder due to better nutrient quality and potential as a cover crop in oil palm plantation, as indicated by the highest biomass production resulting from a higher number of N-fixing and phosphate-solubilizing bacteria, which can increase available N and P [36]. Indonesia is a big producer of plantation commodities including oil palm, rubber, coffee, cocoa, and coconut. Rather than practicing monoculture, our smallholder farmers typically employ integrated farming systems, which encompass various forms such as agroforestry, agro-fisheries, and agro-livestock [27, 37]. In various farming systems, there is always an integration of different populations that complement each other. Shade trees in an agroforestry system maintain the microclimate of the farm and its surroundings. Specifically, it is necessary for coffee and cocoa which cannot withstand high sunlight exposure. In the agroforestry systems, shade trees protect the coffee and cocoa trees from increased air temperatures. The trees under shade experience less stress than those with no shade. Litter leaves of shade trees in coffee agroforestry systems provide additional nutrition for the soil and maintain soil humidity, particularly during prolonged drought and when air temperature increases [37].

Green manure and cover crop management are important to sustain agricultural production in Indonesia. The benefits of green manure and cover crops are not only as a nutrient source, but also for soil organic carbon sequestration, climate change mitigation, and restoring land degradation. Through knowledge accumulation and adaptive management, Indonesian farmers utilize several livelihood adaptation strategies to reduce risks in the face of changing climate conditions. In addition, farmers utilize on-farm crops to obtain various ecosystem services that provide services such as food, nutrition, and medicine and improve ecology, nutrient cycling, and climate regulation [27].

3.5 Biofertilizer

Biofertilizer is an environmentally friendly solution to maintain soil quality. Biological fertilizer is a substance or material that contains live microorganisms (bacteria, fungi, algae) that have the ability to improve or increase soil fertility. Biological fertilizers are able to provide nutrients for plants through natural processes of microorganism activity in the soil such as N fixation, P dissolution, and stimulating plant growth and the synthesis of growth-promoting substances. The microorganisms contained in biological fertilizers can easily convert complex organic materials into simpler compounds so that they can be easily absorbed by plants. Biofertilizers are classified based on the group of microorganisms contained therein, namely biological fertilizers containing N-fixing bacteria (NFB), phosphorus biofertilizers consisting of PSB (P-solubilizing bacteria) and PMB (P-mobilizing bacteria), plant growth booster biological fertilizers (PGPB: plant growth-promoting biofertilizers), K biofertilizers, zinc-solubilizing biofertilizers (ZSB), sulfur-oxidizing biofertilizers (SOB), and silica solvent biofertilizers (SSB) [38].

Microorganisms capable of fixing N are divided into symbiotic and non-symbiotic. Most of the symbiotic N-fixing microorganisms come from the Rhizobiaceae family, which forms mutualistic relationships with legume plants. Meanwhile, non-symbiotic N-fixing microorganisms are carried out by free-living endophytic microorganisms such as *Azotobacter* sp., *Azospirillum* sp., *Cyanobacteria* sp. Nutrients in the soil such as N cannot be absorbed directly by plants, these nutrients require the help of bacteria to fix nitrogen into the form of NH_4^+ and NO_3^- so that it can be absorbed by plants [39].

The implementation of the use of biological fertilizers in Indonesia is increasingly widespread and diverse in line with increasing awareness of the importance of sustainable agriculture. Application of biofertilizer to oil palm plants can increase soil microbial biodiversity and the resistance of beneficial bacteria in the soil [40]. Biofertilizers containing beneficial bacteria improve soil health, supporting plant growth and productivity. In oil palm plantations, biological fertilizers can increase fertilizer efficiency by reducing the need for inorganic fertilizers while maintaining plant growth and bunch weight. Using biological fertilizers in oil palm plantations helps reduce costs and improve the efficiency of chemical fertilizer use. Applying bacteria with nitrogen-fixing, phosphate-solubilizing, potassium-solubilizing, and IAA-producing activities is an environmentally friendly approach to increasing nutrient availability in the soil, thereby enhancing plant growth and development [38, 40]. Simarmata et al. [41] show that the application of a combination of Trichoderma biofertilizer and straw compost can increase the induction of plant resistance (ISR: Induction Systemic Resistance) and productivity in rice plants. The use of biological fertilizer can also reduce the intensity of diseases in rice plants such as brown spot (*Helminthosporium oryzae*), narrow brown spot (*Cercospora janseana*), sheath rice blight (*Rhizoctonia solani*), and bacterial leaf blight diseases (*Xanthomonas oryzae*). Trichoderma biofertilizer and straw compost can increase the number and biodiversity of microorganisms in the soil and improve the plant resistance induction system. The plant resistance induction system has emerged as an important mechanism to promote plant growth and protect plants from pathogen attack. Supriyono et al. [42] reported that the application of biological fertilizer affects the growth of hybrid corn plants in the vegetative phase. The treatment of giving biological fertilizer resulted in higher plant height, stem diameter, and leaf area index of hybrid corn compared to the treatment without giving biological fertilizer. Apart from that, this is because the content of biological fertilizer can help the decomposition process so that the nutrients in the soil can be absorbed by the corn plants. Additionally, biological fertilizer contains microbes including *Azospirillum* sp., *Azotobacter* sp., and *Bacillus* sp. so that it can provide nutrients to improve the growth and development process of hybrid corn plants. Based on Wilujeng et al. [43], the combination of *Azotobacter* biofertilizer and NPK fertilizer can increase vegetative growth in SD3 IPB sweet corn plants, biofertilizer containing *Azotobacter* is able to synthesize the growth hormone IAA (Indole-3-Acetic Acid) or natural auxin which is involved in the regulation of plant growth and development, *Azotobacter* also plays an important role in soil fertility [44].

One of the biological fertilizers that is now increasingly in demand by Indonesian farmers is MOL (local microorganisms). MOL is a microorganism that is usually used as a starter in making both solid and liquid organic fertilizers. The main contents of MOL are carbohydrates, glucose, and sources of microorganisms. Mol can be called a bioactivator which aims to speed up the fermentation process. MOL solution contains various microorganisms such as bacteria and fungi that live together and synergize with each other. The basic ingredients that can be used for MOL fermentation can come from agriculture, plantations, and waste produced from households [45]. MOL is usually made from a mixture of rice washing water and Moringa leaves. Moringa leaves contain cytokinin which plays a role in cell division and accelerates the growth of shoots and stems. Suhastyo and Raditya [46] showed that the concentration and frequency of giving moles of Moringa leaves to pagoda mustard greens (*Brassica narinosa*) had a significant effect on plant growth (plant height), specifically at a frequency of giving once every 6 days with a concentration of 70 ml l^{-1} MOL Moringa leaves. It is believed that the microorganism content in Moringa leaf MOL applied to the soil can stimulate the decomposition process of the planting medium, thereby helping to provide nutrients from organic material available in the soil, which ultimately can increase nutrient absorption by plants, so that plants grow more optimally. Moringa leaves contain high levels of protein, which can be seen from the easy wilting and damage of the leaves, which indicates that the N content is also high.

In Indonesia, *jakaba* or "*eternal lucky fungi*" is one of the fungi that has the potential to be used as a biofertilizer. *Jakaba* is a fungus that grows from the incubation process of water used to wash rice, known as rice washing water. Jakaba was discovered accidentally by farmers in 2016 while producing liquid organic fertilizer. By incubating rice washing water for 14 days, a brown, coral-shaped fungus emerged [47]. Apart from rice washing water, jakaba can also be made by adding bamboo roots. The application of *jakaba* liquid organic fertilizer to corn showed significant differences between treatments. Treatment with a dose of 40 ml l^{-1} *jakaba* liquid media gave the best results on corn growth [48]. This finding is in line with previous research which showed the effectiveness of 40 ml l^{-1} jakaba in increasing leaf length, plant height, and fresh weight of pakcoy (*Brassica rapa* L.) plants [49].

Soil fertility management should consider physical, chemical, and biological aspects to ensure soil health and nutrient availability for plants, which in turn affects sustainable productivity. To minimize environmental damage, the use of eco-friendly materials like biological fertilizers is recommended.

3.6 Biochar

One approach to improving soil quality is through the use of biochar. Biochar is a carbon-rich material produced by burning biomass in a closed container through thermal decomposition with limited oxygen supply and relatively low temperatures (<700°C) [50]. In this context, biochar is an organic material that has been burned and intentionally applied to the land to enhance soil properties. The raw materials for biochar are relatively inexpensive and can be sourced from regions producing agricultural and forestry commodities. Typically, these raw materials include agricultural and forestry biomass residues such as wood chips, coconut shells, bamboo, oil palm fronds, corn cobs, rice husks, peanut shells, and wood processing waste. Biochar has been shown to persist in the soil for over 1000 years and can sequester carbon in the soil [51]. Its application has a positive impact on cultivated plants.

The use of biochar in plant cultivation positively affects productivity and soil quality. Applying biochar in cultivation activities has been shown to increase plant productivity on sandy or acidic soils, and chemically, it improves soil quality and nutrient cycle efficiency [52]. For leafy vegetable crops, biochar application results in increased total biomass, root biomass, plant height, and leaf number, as well as changes in soil physical structure and modification of soil chemical properties over three planting cycles [53]. An illustration of biochar application as a soil amendment and its support for sustainability can be seen in **Figure 1** [54].

The utilization of biochar on dry land in Indonesia has demonstrated positive impacts. Benefits include increasing soil pH and cation exchange capacity (CEC), enhancing soil's water and nutrient retention capacity, and boosting total soil carbon content. Additionally, biochar can reduce CO_2 emission rates, remain stable (resistant to decomposition) in the soil, persist for over 400 years, and serve as a carbon conservation method. It also creates a favorable habitat for microorganisms [55]. Beneficial

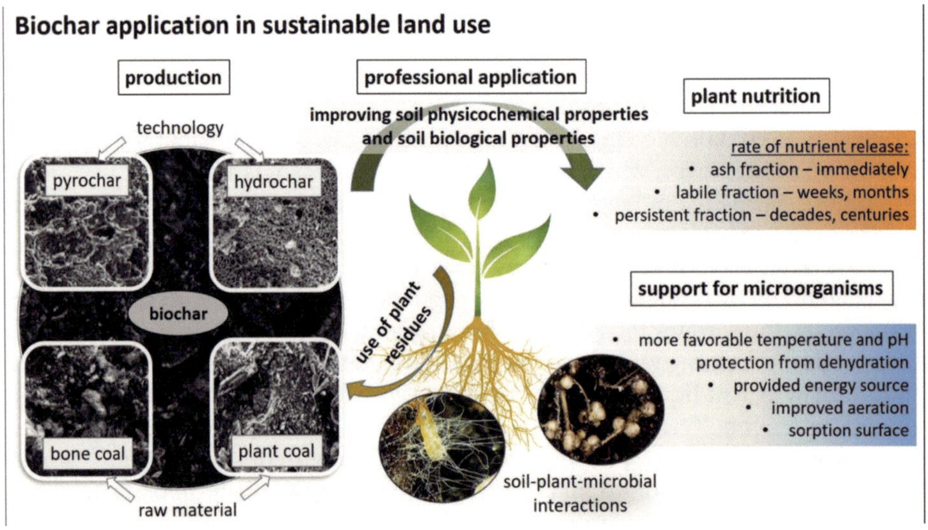

Figure 1.
Types of biochar and their applications in agriculture.

microbes have been found colonizing the surface of biochar, capable of surviving in various environmental conditions. The number of these colonizations increases when biochar is applied to the soil [56]. In heavy metal-contaminated environments, biochar shows significant potential for soil remediation [57].

Biochar can be applied to cultivated land in several ways. It can be spread on the soil surface and then mixed into the soil through tillage. However, surface application has the drawback of biochar being washed away during heavy rain. Alternatively, biochar can be placed in planting furrows, which are then covered with soil, or in planting holes. However, these methods require relatively high labor input [55].

4. Indigenous knowledge related to organic farming and organic fertilizer in Indonesia

Organic farming is actually closely related to the culture of various tribes and customs in Indonesia. Almost every indigenous community in Indonesia has a close relationship with the context of organic farming, although the terminology used varies greatly. According to Schreer et al. [58], the terminology "organic farming" in Indonesia was introduced by non-state actors from abroad, in fact, the organic approach itself has been carried out for a long time.

The understanding that nature will provide food for humans through its mechanisms has long been adopted by various farming communities. One example is the widespread understanding that the land around volcanoes is fertile land, especially after a volcanic eruption. Achmad et al. [59] conducted research on rubber plantations in East Java, which is one of the scientific proofs of this belief in soil fertility. This fertile land is then planted with various agricultural commodities without the addition of ameliorants [60].

Customary rules and traditions are part of the soil and water conservation process, which is translated into agricultural methods that maintain ecosystem balance [61]. Local wisdom passed down to the current generation in several indigenous communities in Indonesia shows how flexible customary rules are in providing a framework for organic farming. *The Kasepuhan Ciptagelar* Indigenous Community in Sukabumi, for example, manages agricultural land and water, and uses polyculture, organic fertilizers, and hereditary seeds in its food supply process [62]. Not only does it regulate the use and system of agriculture, but local wisdom also provides an ethical basis for the communities involved in it to continue to strive for the agricultural system that we currently know as organic farming system. This is clearly reflected, for example, in *the Samin* Indigenous Community [63] which upholds environmental ethics in farming activities, the farming community in Rejang Lebong (Sumatra) which strives for sustainable agriculture [64], *the Dayak Meratus* Community (Kalimantan) with the manugal tradition [65], to the *sasi* system which is still widely practiced by indigenous communities in eastern Indonesia [66].

Sociologically, the social system built by the community also plays a significant role in organic farming practices. For example, Scorviana et al. [67] stated that indigenous women also participate in agricultural cultivation activities. Women tend to carry out agricultural practices by adding organic materials from their kitchen waste as ameliorants to "maintain and increase production yields". This approach is also part of the study of ecofeminism which focuses on equality between women and men in access to natural resources, especially those related to ecosystem sustainability [68].

5. Technology adoption

The use of environmentally friendly organic fertilizers is crucial, especially in sustainable agriculture. The current condition of agricultural land, which is increasingly degraded and polluted, threatens future food production if not addressed. Severely degraded and critical land covers about 48.3 million hectares, or 25.1% of Indonesia's area [69]. During the Green Revolution, the government's pursuit of food self-sufficiency led to extensive use of inorganic fertilizers, resulting in a decline in soil fertility and environmental damage. This reliance also made farmers dependent on inorganic fertilizers [70]. Therefore, agricultural practices must consider ecological limitations. One solution to environmental problems is adopting organic fertilizers, which reduce dependence on chemical fertilizers that leave a harmful environmental footprint.

National Geographic Indonesia reports that inorganic fertilizers contribute 5% of carbon emissions [71]. The rate of use of renewable resources should not exceed their regeneration rate [72]. Currently, continuous use of inorganic fertilizers hardens the soil and depletes nutrients, harming both the environment and farmers in the long term. Organic fertilizers, often derived from organic waste such as food scraps, animal waste, and biomass, contribute to a circular agricultural system by converting pollutants into productive agricultural inputs.

The challenge is why farmers still depend on inorganic fertilizers. According to Rogers' theory of innovation diffusion [73], several factors influence this: the characteristics of the innovation (organic fertilizer), the characteristics of the adopters (farmers), and the communication channels. Organic fertilizers must offer relative advantages over inorganic fertilizers, align with existing agricultural practices and values, and be easy to apply. They should be trialable on a small scale, with visible results for other farmers to see. Farmers who can be encouraged to use organic fertilizer are categorized as innovators and early adopters.

The adoption of organic fertilizers is a complex process from economic, ecological, and social perspectives. By understanding ecological economic theories and innovation diffusion, the government and other stakeholders can provide motivation, policy support, education, and economic incentives to promote more environmentally friendly and sustainable agricultural practices that align with societal values.

6. Production and distribution of organic fertilizer

The production and distribution of organic fertilizers in Indonesia are gaining significant attention as key elements in promoting sustainable agriculture. This growing emphasis stems from the need to improve soil health, enhance crop productivity, and mitigate the adverse environmental effects of chemical fertilizers. Organic fertilizers, sourced from natural materials like plant and animal waste, are becoming an integral part of agricultural systems across the country, reflecting a heightened awareness of their advantages.

In Indonesia, organic fertilizer production commonly utilizes agricultural by-products such as rice husks, animal manure, and compost. The distribution of organic fertilizers in Indonesia is supported by networks of local producers and agricultural cooperatives, ensuring that farmers in remote areas have adequate access to these crucial inputs. By using locally sourced materials, producers lower costs, stimulate local economies, and promote sustainability [74]. Government initiatives, including subsidies and training programs, further encourage farmers to adopt organic fertilizers, fostering the shift toward eco-friendly farming practices [75].

However, challenges persist in the organic fertilizer industry, particularly in processing infrastructure and quality control. Many farmers still rely on traditional production methods, which can result in inconsistent nutrient content and variable effectiveness [76]. Establishing standardized processes and quality assurance systems is essential to build trust and ensure efficacy. Enhanced farmer education on the benefits and proper application of organic fertilizers is equally important to maximize their impact [76].

Economically, organic fertilizer adoption offers long-term benefits for farmers. Reduced reliance on expensive chemical fertilizers leads to lower input costs [77]. Furthermore, the rising demand for organic produce provides farmers with opportunities to access premium markets, creating incentives for sustainable practices [74]. Cooperative models for organic fertilizer production can enhance profitability for smallholder farmers by fostering resource sharing and collective knowledge [78].

Beyond individual farms, the widespread adoption of organic fertilizers contributes to national food security. By building resilient agricultural systems less dependent on synthetic inputs, Indonesia can better adapt to climate change and market fluctuations while advancing its agricultural development goals [79]. This transition supports the dual objectives of increasing productivity and protecting the environment in Indonesia [76].

Looking ahead, the future of organic fertilizer production in Indonesia is promising. Research and innovation are driving improvements in formulations and application techniques. For instance, bio-organic fertilizers, which blend organic matter with beneficial microorganisms, are gaining momentum due to their ability to boost yields and enhance soil health. Such advancements underscore the critical role of organic fertilizers in advancing sustainable farming and securing agricultural future in Indonesia.

7. Conclusion

Organic fertilizers are essential for sustainable agricultural production in Indonesia, having been used in natural farming practices long before the country was established. These fertilizers come in various forms, including compost (from crop residues, manure, vermicompost, and maggot compost), green manure, biofertilizers, and biochar. Compost and animal manure are the most common types of organic fertilizers. Compared to conventional compost, vermicompost using earthworms accelerates decomposition, eliminates odors, and produces high-quality compost. Similarly, bioconversion mediated by BSF larvae (Maggcompost) effectively transforms organic waste within a day, enabling high-capacity production. Maggcompost contains chitin, microorganisms, and bio-stimulants that enhance soil fertility, suppress pathogens, and strengthen plant resistance. Another approach to increasing organic matter in the soil involves using green manure or cover crops in farming rotation. Green manure and cover crops serve as nutrient sources, facilitate soil organic carbon sequestration, mitigate climate change, and restore land degradation. Additionally, biofertilizers containing living microorganisms improve soil fertility by providing nutrients through natural soil processes such as nitrogen fixation, phosphate dissolution, stimulating plant growth, and synthesizing growth-promoting substances. Biochar amendments in plant cultivation positively affect productivity and soil quality. Utilizing biochar on dry land in Indonesia enhances soil pH, cation exchange capacity (CEC), water and nutrient retention, and total soil carbon content. It reduces CO_2 emissions, remains stable in the soil, supports microorganisms, and shows potential for soil remediation in heavy metal-contaminated environments. Various organic fertilizers are primarily used as soil amendments to enhance soil and plant health. They can be combined with inorganic fertilizers or applied alone. Effective organic farming requires managing organic fertilizers in conjunction with other practices to boost crop productivity and economic value. The long tradition of organic farming and fertilizers is maintained by local communities. The future of organic fertilizer production in Indonesia is promising but challenging. Local producers and agricultural cooperatives support production and distribution, ensuring farmers in remote areas have access to these essential inputs. Government initiatives, including subsidies and training programs, further encourage farmers to adopt organic fertilizers, fostering the shift toward eco-friendly farming practices and enhancing agricultural sustainability.

Acknowledgements

This work was supported by the College of Vocational Studies, IPB University.

Conflict of interest

The authors declare no conflict of interest.

Author details

Ratih Kemala Dewi[1,2*], Restu Puji Mumpuni[1], Muhammad Iqbal Nurulhaq[1], Agief Julio Pratama[1], Edi Wiraguna[1], Ririh Sekar Mardisiwi[1], Widya Hasian Situmeang[1], Tri Budiarto[1] and Henry Kasman Hadi Saputra[1,3]

1 College of Vocational Studies, IPB University, Bogor, Indonesia

2 College of Agriculture, Ibaraki University, Ibaraki, Japan

3 Graduate School of Science and Engineering, Ibaraki University, Ibaraki, Japan

*Address all correspondence to: ratihkemala@apps.ipb.ac.id

References

[1] Indonesian Ministry of agriculture. Statistik Makro Sektor Pertanian 2023. Jakarta: Indonesian Ministry of Agriculture; 2023

[2] Setyanto SR. Wulang Tani, Edisi Teks dan Gambaran Tata Cara Pertanian. Jakarta, Indonesia: Perpustakaan Nasional RI; 2018. 128 p

[3] Paull J. A century of synthetic fertilizer: 1909-2009. Elements Journal of Bio-Dynamics Tasmania [Internet]. 2009;**94**:16-21. Available from: http://orgprints.org/15797/01/15797.pdf

[4] Paull J. Attending the first organic agriculture course: Rudolf Steiner's agriculture course at Koberwitz, 1924. European Journal of Social Sciences. 2011;**21**(1):64-70

[5] Wilis L. Pertanian Biodinamik: Studi Kasus Sistem Produksi Anggur Alternatif di Ihringen, Jerman Selatan. Lembaran Antropologi. 2022;**1**(1):83-98

[6] Kirchmann H. Biological dynamic farming - an occult form of alternative agriculture? Journal of Agricultural and Environmental Ethics. 1994;7(2):173-187

[7] Mahbub IA, Tampubolon G, Mukhsin M, Farni Y. Peningkatan kesuburan tanah dan hasil padi sawah melalui aplikasi pupuk organik. Jurnal Tanah dan Sumberdaya Lahan. 2023;**10**(2):335-340

[8] Moe K, Moh SM, Htwe AZ, Kajihara Y, Yamakawa T. Effects of integrated organic and inorganic fertilizers on yield and growth parameters of rice varieties. Rice Science. 2019;**26**(5):309-318

[9] Gusmiatun MN. Peran pupuk organik dalam mengurangi pupuk anorganik pada budidaya padi gogo (role in reducing organic fertilizers inorganic fertilizer on rice culture upland). Jurnal Ilmiah Agribisnis dan Perikanan (agrikan UMMU-Ternate). 2018;**11**(2):91-99

[10] Razavipour T, Moghaddam SS, Doaei S, Noorhosseini SA, Damalas CA. Azolla (*Azolla filiculoides*) compost improves grain yield of rice (*Oryza sativa* L.) under different irrigation regimes. Agricultural Water Management [Internet]. 2018;**209**:1-10. DOI: 10.1016/j.agwat.2018.05.020

[11] Hoque TS, Hasan AK, Hasan MA, Nahar N, Dey DK, Mia S, et al. Nutrient release from vermicompost under anaerobic conditions in two contrasting soils of Bangladesh and its effect on wetland rice crop. Agriculture. 2022;**12**(3):1-17

[12] Bachtiar T, Robifahmi N, Flatian AN, Slamet S, Citraresmini A. Pengaruh dan kontribusi pupuk kandang terhadap n total, serapan n (15n), dan hasil padi sawah (*Oryazae sativa* l.) varietas Mira-1. Jurnal Sains dan Teknologi Nuklir Indonesia. 2020;**21**(1):35

[13] Wu YP, Shaaban M, Deng CJ, Peng QA, Hu RG. Changes in the soil N potential mineralization and nitrification in a rice paddy after 20 yr application of chemical fertilizers and organic matter. Canadian Journal of Soil Science. 2017;**97**(2):290-299

[14] Taheri Rahimabadi E, Ansari MH, Razavi NA. Influence of cow manure and its vermicomposting on the improvement of grain yield and quality of rice (*Oryza sativa* L.) in field conditions. Applied Ecology and Environmental Research. 2018;**16**(1):97-110

[15] Sastro Y. Teknologi Vermikomposting Limbah Organik Kompos. Jakarta, Indonesia: Balai Pengkajian Teknologi Pertanian (BPTP); 2016. pp. 1-35. Available from: http://repository.pertanian.go.id/handle/123456789/8738

[16] Norbu T. Pretreatment of Municipal Solid Waste by Windrow Composting and Vermicomposting [Internet] [thesis]. Pathumthani, Thailand: Asian Institute of Technology; 2002. Available from: http://www.faculty.ait.ac.th/visu/data/

[17] Singh RP, Singh P, Araujo AS, Ibrahim MH, Sulaiman O. Management of urban solid waste: Vermicomposting a sustainable option. Resources, Conservation and Recycling. 2011;**55**(7):719-729

[18] Ghosh M, Chattopadhyay GN, Baral K. Transformation of phosphorus during vermicomposting. Bioresource Technology. 1999;**69**(2):149-154

[19] Suthar S. Vermicomposting of vegetable-market solid waste using *Eisenia fetida*: Impact of bulking material on earthworm growth and decomposition rate. Ecological Engineering. 2009;**35**(5):914-920

[20] Hammermeister AM, Astatkie T, Jeliazkova EA, Warman PR, Martin RC. Nutrient supply from organic amendments applied to unvegetated soil, lettuce and orchardgrass. Canadian Journal of Soil Science. 2006;**86**(1):21-33

[21] Atiyeh RM, Subler S, Edwards CA, Bachman G, Metzger JD, Shuster W. Effects of vermicomposts and composts on plant growth in horticultural container media and soil. Pedobiologia (Jena). 2000;**44**(5):579-590

[22] Sakhidin, Widarawati R, Naila Khusna Syarifah R, Prakoso B, Hanifa H, Widyasunu P, et al. Teknologi Pertanian Dalam Mendukung Pertanian Yang Maju Dan Berkelanjutan. Indonesia: Universitas Jendral Soedirman; 2024

[23] Abd Manan F, Yeoh YK, Chai TT, Wong FC. Unlocking the potential of black soldier fly frass as a sustainable organic fertilizer: A review of recent studies. Journal of Environmental Management [Internet]. 2024;**367**(121997):1-11. DOI: 10.1016/j.jenvman.2024.121997

[24] Elissen H, Schilder M, Postma J, Weide R van der. Disease Suppressing Effects of BSF (Black Soldier Fly, *Hermetia illucens*) Larvae Frass in Bioassays with Cress and Sugar Beet Plants [Internet]. Wageningen: Wageningen University & Research; 2019. Available from: https://core.ac.uk/download/pdf/287784851.pdf

[25] Choi S, Hassanzadeh N. BSFL frass: A novel biofertilizer for improving plant health while minimizing environmental impact. The Canadian Science Fair Journal [Internet]. 2019;**2**(2):41-46. Available from: https://csfjournal.com/s/Sarah-and-Neelah-mjep.pdf

[26] Agustiyani D, Agandi R, Arinafril NAA, Antonius S. The effect of application of compost and frass from black soldier fly larvae (*Hermetia illucens* L.) on growth of Pakchoi (*Brassica rapa* L.). IOP Conf Ser Earth Environ Sci. 2021;**762**(012036):1-10

[27] Yusriadi Y, Cahaya A, Hamzah F. Farmer adaptation strategies through local farming systems in Enrekang, Indonesia. Scientific Reports. 2024;**14**(21652):1-14

[28] Ali M. Evaluation of green manure technology in tropical lowland rice systems. Field Crops Research. 1999;**61**(1):61-78

[29] Sakiah SM, Hasibuan J. Entisol land characteristics with and without cover crop (*Mucuna bracteata*) on rubber plantation. IOP Conference Series: Earth and Environmental Science. 2018;**122**(012043):1-4

[30] Brewbaker JL, Glover N. Woody species in rice-based cropping systems. In: Green Manure in Rice Farming: The Role of Green Manure Crops in Rice Farming System. Manila, Philippines: International Rice Research Institute; 1988. pp. 29-43

[31] Setiawati MR, Suryatmana P, Budiasih SN, Nurlina L, Kurnani BA, et al. Utilization *Azolla pinnata* as substitution of manure to improve organic rice yield and paddy soil health. IOP Conference Series: Earth and Environmental Science. 2018;**215**:012006

[32] Carangal VR, Rebancos ET, Armada EC, Tengco PL. Integration of food, forage, and green manure production systems. In: Ladha JK, Garrity DP, editors. Selected Papers from International Rice Research Conference: Green Manure Production Systems for Asian Ricelands. Manila, Philippines: International Rice Research Institute; 1994. pp. 52-65

[33] Subaedah S, Aladin A, Nirwana. Fertilization of nitrogen, phosphor and application of green manure of crotalaria *Juncea* in increasing yield of maize in marginal dry land. Agriculture and Agricultural Science Procedia [Internet]. 2016;**9**:20-25. DOI: 10.1016/j.aaspro.2016.02.114

[34] Idham I, Pagiu S, Lasmini SA, Nasir BH. Effect of doses of green manure from different sources on growth and yield of maize in dryland. International Journal of Design & Nature and Ecodynamics. 2021;**16**(1):61-67

[35] Comte I, Colin F, Whalen JK, Grünberger O, Caliman JP. Agricultural practices in oil palm plantations and their impact on hydrological changes, nutrient fluxes and water quality in Indonesia. A review. Advances in Agronomy. 2012;**116**:71-124

[36] Hutasoit R. Evaluation of four pasture legumes species as forages and cover crops in oil palm plantation. Jurnal Ilmu Ternak dan Veteriner. 2018;**22**(3):124-134

[37] Martini E, Riyandoko RJM. In: Finlayson RF, editor. Establishing Coffee-Agroforestry Systems [Internet]. Bogor, Indonesia: World Agroforestry Centre (ICRAF) Southeast Asia Regional Program; 2017. 156 p. Available from: https://library.sweetmarias.com/wp-content/uploads/2020/08/GUIDELINES-FOR-Establishing-Coffee-Agroforestry-Systems.pdf

[38] Itelima JU, Bang WJ, Onyimba IA, Sila MD, Egbere OJ. Bio-fertilizers as key player in enhancing soil fertility and crop productivity: A review. Journal of Microbiology [Internet]. 2018;**2**(1):74-83. Available from: http://irepos.unijos.edu.ng/jspui/handle/123456789/1999%0Ahttps://dspace.unijos.edu.ng/jspui/bitstream/123456789/1999/1/Itelima-et-al%281%29.pdf

[39] Sneha, Anitha, Sahair A, Nataraj R, Thangaraj G, G.K C, et al. Biofertilizer for crop production and soil fertility. Academia Journal of Agricultural Research [Internet]. 2018;**6**(8):299-306. Available from: https://www.researchgate.net/publication/332301492

[40] Hidayat F, Yudhistira Y, Desika Putri Pane R, Sapalina F, Listia E, Amalia R, et al. Aplikasi Pupuk Hayati Untuk Meningkatkan Pertumbuhan dan Produktivitas Tanaman Kelapa

Sawit. Jurnal Penelitian Kelapa Sawit. 2023;**31**(2):96-107

[41] Simarmata T, Hersanti TT, Fitriatin BN, Setiawati MR, Purwanto. Application of bioameliorant and biofertilizers to increase the soil health and rice productivity. HAYATI Journal of Biosciences [Internet]. 2016;**23**:181-184. DOI: 10.1016/j.hjb.2017.01.001

[42] Supriyono S, Nurmalasari AI, Sulistyo TD, Fatimah S. Efektivitas Pupuk Hayati terhadap Pertumbuhan dan Hasil Jagung Hibrida di Tanah Alfisol. Agrotechnology Research Journal. 2022;**6**(1):1-7

[43] Wilujeng EDI, Ismiani S, Adnan MR. Respon pertumbuhan tanaman jagug manis varietas SD3 IPB terhadap pemberian kombinasi NPK dan pupuk hayati Azotobacter. Gontor Agrotech Science Journal. 2022;**8**(2):102-108

[44] Jnawali AD, Ojha RB, Marahatta S. Role of Azotobacter in soil fertility and sustainability–a review. Advances in Plants & Agriculture Research. 2015;**2**(6):1-5

[45] Ekawandani N, Halimah N. Pengaruh Penambahan Mikroorganisme Lokal (MOL) dari Nasi Basi Terhadap Pupuk Organik Cair Cangkang Telur. Biosfer: Jurnal Biologi dan Pendidikan Biologi. 2021;**6**(2):2-9

[46] Suhastyo AA, Raditya FT. Respon pertumbuhan dan hasil sawi pagoda (*Brassica narinosa*) terhadap pemberian mol daun kelor. Agrotechnology Res J. 2019;**3**(1):56-60

[47] Ani KA, Ernawati E, Suryani S, Adonis E, Manao A, Ariyana R. Budidaya Jakaba dan Aplikasi sebagai Insektisida terhadap Larva Ostrinia fulnacaris Guenee. Vol. 9. Bioeksperimen: Jurnal Penelitian Biologi; 2023. pp. 42-49

[48] Fadilah RA, Kurnia MD, Putra IP. Jakaba undercover: Taxonomic riddle and potency in Indonesian agriculture. Caraka Tani: Journal of Sustainable Agriculture. 2024;**39**(2):411-423

[49] Apriyanto A, Ibnusina F, Afrizal R. Pemberian Dosis POC Jakaba Terhadap Pertumbuhan dan Produksi Tanaman Pakcoy (*Brassica rapa* L.). Perbal: Jurnal Pertanian Berkelanjutan. 2023;**11**(3):343-351

[50] Lehmann J, Joseph S. Biochar for Environmental Management: Science and Technology. London, UK: Earthscan; 2009. 438 p

[51] Lehmann J. Sequestration of carbon in tropical forests A concept paper for PNG. Nature. 2007;**447**:143-144

[52] Dai Y, Zheng H, Jiang Z, Xing B. Combined effects of biochar properties and soil conditions on plant growth: A meta-analysis. Science of the Total Environment [Internet]. 2020;**713**(136635):1-43. Available from: https://www.sciencedirect.com/science/article/abs/pii/S0048969720301455

[53] Carter S, Shackley S, Sohi S, Suy TB, Haefele S. The impact of biochar application on soil properties and plant growth of pot grown lettuce (*Lactuca sativa*) and cabbage (*Brassica chinensis*). Agronomy. 2013;**3**(2):404-418

[54] Kocsis T, Ringer M, Biró B. Characteristics and applications of biochar in soil–plant systems: A short review of benefits and potential drawbacks. Applied Sciences. 2022;**12**(8):1-15

[55] Nurida NL, Achmad R, Sutono S. In: Soelaeman Y, Purnomo J, editors. Biochar: Pembenah Tanah Yang Potensial. Jakarta, Indonesia: IAARD Press; 2015. 48 p

[56] Zhang M, Xia H, Riaz M, Bo L, Zeinab-El-Desouki JC. Various beneficial microorganisms colonizing on the surface of biochar primarily originated from the storage environment rather than soil environment. Applied Soil Ecology. 2023;**182**:104700

[57] Wu J, Fu X, Zhao L, Lv J, Lv S, Shang J, et al. Biochar as a partner of plants and beneficial microorganisms to assist in-situ bioremediation of heavy metal contaminated soil. Science of the Total Environment. 2024;**923**:171442

[58] Schreer V, Padmanabhan M. The many meanings of organic farming: Framing food security and food sovereignty in Indonesia. Organic Agriculture. 2020;**10**(3):327-338

[59] Achmad SR, Hadi H. Identification of chemical properties of volcanic ash and the effort of recovery of rubber plants affected by eruption of Kelud Mountain (case study: Ngrangkah Pawon field, East Java). Warta Perkaretan. 2015;**34**(1):19-30

[60] Zaenurrohman JA, Qur'an MIN, Ismangil, Siswandi, Candra A. Volcanic rock of slamet volcano as the potential of soil ameliorant. Indian Journal Of Geosciences. 2024;**11**(1):81-90

[61] Prabowo YB, Sudrajat S. Kearifan Lokal Kasepuhan Ciptagelar: Pertanian Sebagai Simbol Budaya & Keselarasan Alam. Jurnal Adat dan Budaya Indonesia. 2021;**3**(1):6-16

[62] Muharam H, Gursida H, Susilowati IH, Widiyanti W, Alawiah ET, Setianingrum HW. Menggali Dan Membangun Keunggulan Kompetitif Melalui Kearifan Lokas Pada Masyarakat Adat Kasepuhan Cipta Gelar Kabupaten Sukabumi. Jurnal Abdi Masyarakat Multidisiplin. 2023;**2**(3):22-26

[63] Kurniawan N, Budiaman B, Hidayah AN, Salsabilla A. Nilai etika lingkungan pada pertanian komunitas adat Sedulur Sikep Samin. Jurnal Pengelolaan Lingkungan Berkelanjutan (JPLB) Journal of Environmental Sustainability Management. 2023;**7**(3):258-267

[64] Kesumawati N, Armadi Y, Hayati R. Pentingnya Menggali Kearifan Lokal Masyarakat Petani Dalam Rangka Memelihara Kelestarian Lahan Pertanian Sebagai Pendukung Ketahanan Pangan Di Kabupaten Rejang Lebong. Agribis Journal [Internet]. 2021;**14**(2):1-10. Available from: http://jurnal.umb.ac.id/index.php/agribis/article/view/1740

[65] Efendi M, Sahrul M, Salma S. Nilai Kearifan Lokal Tradisi Manugal Masyarakat Dayak Meratus Kalimantan Selatan Pada Materi Geografi Bidang Lingkungan Hidup (Kajian Etnografi). PADARINGAN (Jurnal Pendidikan Sosiologi Antropologi). 2020;**2**(2):260-270

[66] Wahyudin Y, Wahyudin M. Mengenal Hukum Adat "Larwul Ngabal" Masyarakat Kepuluan Kei Maluku Tenggara (knowing customary law "Larwul Ngabal" of the Kei Islands Society in the Southeast Maluku District). SSRN Electronic Journal. 2012;**3**(3):1-25

[67] Scorviana HN, Siscawati M. Strategi Perempuan Adat Pendatang dalam Merespon Sistem Budaya Padi sebagai Sumber Penghidupan. Journal of Education for Sustainable Development. 2023;**24**(01):14-27

[68] Situmeang WH, Aflaha FR. Ragam Modal Perempuan Perdesaan dalam Menghadapi Perubahan Iklim di Tengah Subordinasi. Jurnal Perempuan. 2022;**27**(3):241-253

[69] Wahyunto DA. Degradasi Lahan di Indonesia: Kondisi Existing, Karakteristik, dan Penyeragaman Definisi Mendukung Gerakan Menuju Satu Peta. Jurnal Sumberdaya Lahan. 2014;**8**(2):81-93

[70] Gultom F, Harianto S. Revolusi Hijau Merubah Sosial-Ekonomi Masyarakat Petani. TEMALI: Jurnal Pembangunan Sosial. 2021;**4**(2):145-154

[71] Jenihansen R. Melebihi Industri Penerbangan, Pupuk Menyumbang 5 Persen Emisi Karbon [Internet]. Jakarta: Nationalgeographic; 2023. Available from: https://nationalgeographic.grid.id/read/133690604/melebihi-industri-penerbangan-pupuk-menyumbang-5-persen-emisi-karbon?page=all [Accessed: November 26, 2024]

[72] Daly HE. Ecological Economics and Sustainable Development, Selected Essays of Herman Daly [Internet]. Northampton, Massachussetts, USA: Edward Elgar Publishing; 2008. 288 p. Available from: https://www.e-elgar.com/shop/gbp/ecological-economics-and-sustainable-development-selected-essays-of-herman-daly-9781847209887.html

[73] Rogers EM. Diffusion of Innovations. 5th ed. New York, USA: Free Press; 1995. 519 p

[74] Ciptono EB, Khoir M. Composition analysis of Bokashi organic fertilizer from fish flour fishery waste. AGARICUS: Advances Agriculture Science & Farming. 2022;**2**(1):14-23

[75] Warr P, Yusuf AA. Fertilizer subsidies and food self-sufficiency in Indonesia. Agricultural Economics (United Kingdom). 2014;**45**:1-18

[76] Kurniawan SB. A review of the future of biomass-based fertilizer in Indonesia. EPRA International Journal of Economic and Business Review. 2023;**11**(7):27-31

[77] Atman A, Bakrie B, Indrasti R. Effect of cow manure dosages as organic fertilizer on the productivity of organic rice in West Sumatra, Indonesia. International Journal of Environment, Agriculture and Biotechnology. 2018;**3**(2):506-511

[78] Achmad B, Sanudin B, Siarudin M, Widiyanto A, Diniyati D, Sudomo A, et al. Traditional subsistence farming of smallholder agroforestry systems in Indonesia: A review. Sustainability. 2022;**14**(8361):1-33

[79] Mujiyana M, Islami N, Setiawan NC. Utilization of bio-pesticides as a solution to farmer dependence on the use of chemical pesticides in Sawahan Village, Donorejo District, Pacitan regency. Proceeding International Conference of Technology on Community and Environmental Development. 2023;**1**(2):462-469

Chapter 8

Sustainable Intensification: Agroecosystem and Plant Nutrient Management in a Changing Climate

Kwaku Onwona-Hwesofour Asante, Jeannette Aduhene-Chinbuah, Clement Oppong Peprah, Kennedy Agyeman, Felix Frimpong, Eric Owusu Danquah and Ralph Bam

Abstract

This book chapter addresses the urgent need to shift from traditional practices to sustainable intensification, particularly in regions where environmental constraints and the impacts of climate change are pronounced. It provides an in-depth overview of sustainable intensification, focusing on its core principles to increase food production from existing farmland while minimizing environmental harm. Key aspects discussed include agroecosystem management, where the complex interactions between plants, soil, and microorganisms are explored, along with the critical role of soil biodiversity in maintaining ecosystem productivity. The chapter also emphasizes the importance of proper nutrient management, analyzing the effects of nutrient deficiency on plant health and productivity. It advocates for integrated nutrient management to optimize soil health and crop yields. Innovative practices such as breeding for nutrient use efficiency, crop rotation, intercropping, and advanced technologies like precision agri culture are also highlighted as vital strategies for sustainable intensification. Moreover, the chapter underscores the necessity of supportive policy frameworks and community engagement in promoting sustainable agricultural practices. The importance of sustainable intensification in ensuring food security and economic stability has been thoroughly discussed, offering practical recommendations for stakeholders to foster resilient and sustainable agroecosystems in the face of climate change.

Keywords: sustainable intensification, agroecosystem management, plant nutrient management, tropical agriculture, climate change adaptation

1. Introduction

The global challenge of ensuring food security is intensifying due to a rapidly expanding population and the corresponding surge in food demand. Projections indicate that by 2050, the global population will approach 10 billion, necessitating a substantial increase in food production to meet these demands [1]. This challenge is compounded by environmental constraints that demand adopting sustainable agricultural practices. As the world grapples with feeding an ever-growing population, it becomes crucial to strike a delicate balance between enhancing food production and preserving environmental health [2]. The dual pressures of biodiversity loss and climate change further complicate this balance, posing significant challenges to sustainable food production systems. Biodiversity loss, often driven by agricultural expansion and intensification, undermines ecosystem resilience and their ability to provide essential services, such as pollination and pest control. Climate change, marked by rising temperatures, shifting precipitation patterns, and more frequent extreme weather events, directly threatens crop yields and food security. These interconnected challenges underscore the urgency of adopting sustainable agricultural practices that can adapt to and mitigate these threats.

In response to these challenges, the concept of sustainable intensification (SI) has emerged as a key strategy. Sustainable intensification focuses on increasing agricultural productivity while ensuring environmental sustainability [3]. This approach integrates the dual goals of boosting crop yields and maintaining or improving the ecological systems on which agriculture depends. Sustainable intensification prioritizes optimizing resource use efficiency and reducing the environmental footprint of agriculture, thereby fostering a more harmonious relationship between agricultural practices and the natural world [4]. Within this context, the importance of sustainable nutrient management becomes evident. Effective nutrient management is critical in mitigating the adverse effects of climate change on agriculture [5]. Farmers can enhance soil fertility, increase crop resilience to climate variability, and reduce greenhouse gas emissions by adopting sustainable practices. These practices support higher and more stable food production and contribute to broader environmental objectives, such as carbon sequestration and biodiversity conservation. This chapter explores the principles and practices of sustainable nutrient management within the framework of SI. Through an examination of current research and practical applications, we aim to highlight the importance of integrating productivity with environmental health in modern agriculture.

2. Sustainable intensification

2.1 Historical evolution

Agriculture has historically relied on monoculture cropping, where the same crop is planted continuously, leading to soil nutrient depletion and increased pest vulnerability. This approach, combined with poor soil management, resulted in significant declining soil fertility over time [6]. To counter these issues, sustainable intensification emerged as a critical approach. It focuses on increasing food production from existing farmland while minimizing environmental impact, incorporating innovative practices and technologies to ensure long-term sustainability. Key elements of sustainable intensification include diversification, which enhances crop diversity through intercropping, crop rotation, and mixed farming, improving soil health and resilience to climate variability [7]. Precision agriculture is another vital development, leveraging technology like soil sensors and GPS to optimize resource use, reduce waste, and minimize environmental impact. These advancements are essential for meeting the growing global food demand while maintaining ecological balance [8].

2.2 Key milestones and developments

2.2.1 The Green Revolution and shift toward agroecology

The Green Revolution of the mid-twentieth century, led by figures like Norman Borlaug, introduced high-yielding crop varieties, synthetic fertilizers, and advanced irrigation techniques, significantly boosting food production and preventing famines [9]. However, it also led to environmental and social challenges, such as soil degradation and reduced biodiversity due to intensive chemical use [10]. These issues prompted a shift toward agroecology, which integrates ecological principles into farming, emphasizing sustainability and resilience. This shift was driven by both scientific research and grassroots movements advocating for environmentally sustainable and socially equitable farming practices [11].

2.2.2 The rise of precision agriculture

Precision agriculture (PA) has emerged as a cornerstone of sustainable farming in the twenty-first century. While PA technologies have existed since the 1980s, their adoption was initially slow. By 2012, only about 30–50% of US corn and soybean acres utilized PA, a low figure partly due to the complexity of the required technology and the migration of skilled individuals to urban areas [12]. Despite these challenges, PA adoption varies regionally, as seen in **Table 1**. Technological advancements, including GPS-guided machinery and soil sensors, have enabled data-driven decision-making, improving efficiency and sustainability. For example, PA has been shown to reduce nitrogen use by up to 20% in the United States while maintaining or improving crop yields [20].

Agroecological zone	Region	Precision agriculture tools usage	References
Tropical	Latin America	High usage of drones, satellite imaging, and GPS-guided machinery	[13]
Tropical	Sub-Saharan Africa	Limited usage of precision agriculture tools but growing adoption of mobile apps and SMS-based services	[14]
Temperate	North America	Widespread adoption of precision agriculture tools, including drones, satellite imaging, and precision irrigation	[15]
Temperate	Europe	High usage of precision agriculture tools, including GPS-guided machinery and precision irrigation	[16]
Arid	Middle East and North Africa	Limited usage of precision agriculture tools but growing adoption of drought-tolerant crops and precision irrigation	[17]
Arid	Australia	High usage of precision agriculture tools, including drones and satellite imaging, to manage water resources	[18]
Mountainous	Asia	Limited usage of precision agriculture tools but growing adoption of terracing and contour farming	[19]

Table 1.
Regional variation in precision agriculture tools usage across different agroecological zones.

2.2.3 Integration of sustainable practices, policy support, and community-led initiatives

In recent years, we have witnessed a significant integration of sustainable practices into mainstream agriculture, supported by national and international policies. Examples include the UN's Sustainable Development Goals (SDGs), the Paris Agreement, and the Global Alliance for Climate Smart Agriculture, which emphasize the importance of sustainable practices [21]. National policies, including the European Union's Common Agricultural Policy (CAP) and Brazil's Low Carbon Agriculture Plan, also play a critical role [22]. Community-led initiatives, such as the System of Rice Intensification (SRI) and the "Push-Pull" pest management strategy in East Africa, highlight the effectiveness of involving local communities in agricultural innovation [23]. Initiatives like Participatory Irrigation Management (PIM) in India and community seed banks globally further demonstrate the benefits of local participation in managing resources and preserving biodiversity.

2.3 Core principles

2.3.1 Enhancing productivity with a minimal environmental footprint

One core principle of sustainable intensification is enhancing productivity while minimizing environmental impact. Precision nutrient management, supported by regular soil testing and integrated nutrient management (INM), optimizes nutrient use and improves soil health [24]. Integrated pest management (IPM) combines biological control, cultural practices, and reduced chemical pesticide use to manage pests effectively, protecting non-target organisms and reducing environmental contamination. Conservation agriculture, including minimum tillage and cover cropping, preserves soil structure, reduces erosion, and maintains organic matter, contributing to soil health and productivity. Agroforestry integrates trees with crops, enhancing biodiversity, providing ecosystem services, and improving soil fertility [25].

2.4 Balancing economic, social, and environmental goals

Balancing economic, social, and environmental goals is fundamental to sustainable intensification. Ensuring market access, fair prices, and strengthening value chains for sustainable agricultural products support farmers' livelihoods [26]. Social equity ensures that the benefits of sustainable intensification reach smallholders, women, and marginalized communities. Empowering farmers through knowledge sharing and capacity building is essential for inclusive and equitable agricultural development [27]. Promoting ecosystem services through agroecological practices helps achieve these goals. Economic balance is demonstrated by organic certification programs and Fairtrade initiatives, which provide market access and fair wages, ensuring profitability for farmers [28]. Social goals are further supported by initiatives like Farmer Field Schools, which enhance knowledge sharing and empower marginalized groups, promoting social equity.

3. Agroecosystem management

Agroecosystem management involves applying ecological and agronomic principles to optimize interactions between plants, soil, and microorganisms for sustainable production. This approach is crucial for maintaining ecological balance and enhancing agricultural productivity. Agroecosystems are complex networks of plants, soil, and microorganisms, all of which play vital roles in ecosystem stability and productivity. Biodiversity within these systems is key to promoting resilience and reducing dependency on chemical inputs, contributing to sustainable agricultural practices [29].

3.1 Soil health and biodiversity and agroecosystem services

Soil organisms are integral to nutrient cycling and crucial for soil fertility and plant growth. Conventional farming practices, reliant on chemical inputs and monoculture, often degrade soil health by reducing biodiversity and disrupting nutrient cycling [30]. In contrast, sustainable practices such as crop rotation, organic amendments, and reduced tillage lead to healthier, more productive soils. Agroecosystem services, including pollination, pest control, and nutrient recycling, are vital for sustainable agriculture. Practices like conservation tillage and cover cropping preserve soil structure, reduce erosion, and improve water retention and organic matter [31].

3.2 Integration of agroforestry and permaculture principles

Agroforestry and permaculture integrate ecological principles into farming, enhancing both economic outputs and ecological services. Agroforestry, for instance, provides timber, fruit, and habitat for wildlife, while also contributing to carbon storage and water cycle regulation. Permaculture, on the otherhand, focuses on mimicking natural ecosystems to create sustainable, self-sufficient agricultural systems, emphasizing diversity, stability, and resilience. For instance, in sub-Saharan Africa, the push-pull strategy integrates specific crops to repel pests and attract beneficial insects. This strategy reduces pest damage and increases yields without chemical pesticides, offering a promising path toward sustainable agriculture. Adapting these practices to local contexts through community engagement is crucial for the success and sustainability of agricultural interventions [32, 33].

3.3 Challenges and opportunities

Adopting agroecosystem management practices can be challenging due to limited access to information, financial resources, and understanding of long-term benefits, especially for smallholder farmers [34]. Additionally, inadequate policy frameworks and institutional support often hinder the implementation of sustainable practices. Emerging technologies, such as precision agriculture, satellite imagery, and artificial intelligence (AI), offer opportunities to optimize agroecosystem management, improving planning, resource allocation, and productivity. Enhanced collaboration among researchers, farmers, policymakers, and NGOs can facilitate knowledge transfer, resource sharing, and policy advocacy, fostering innovation and scaling successful practices more effectively.

4. Plant nutrient management in sustainable intensification

Sustainable intensification emphasizes efficient plant nutrient management to boost productivity while minimizing environmental impacts. This approach integrates biodiversity, soil health, and ecological interactions, utilizing practices like cover cropping, integrated crop-livestock systems, and strategic use of legumes to maintain ecological balance and resilience [35]. Macronutrients such as nitrogen, phosphorus, and potassium are critical for plant functions, including growth and resilience. Though needed in smaller quantities, micronutrients are vital for processes like enzyme function and photosynthesis. For example, nitrogen is central to amino acid synthesis, while phosphorus is crucial for energy transfer. Micronutrients like iron, zinc, and manganese play roles in chlorophyll synthesis, enzyme function, and photosynthesis [36]. Nutrient cycling involves the movement and availability of essential nutrients, like nitrogen and phosphorus, which are often limited in agroecosystems. Processes like nitrogen fixation by legumes and phosphorus mobilization by soil microorganisms enhance nutrient availability. However, nutrients can be lost through leaching, erosion, and gaseous emissions, necessitating careful soil fertility management [37].

Nutrient deficiencies and abiotic stresses, such as drought and salinity, negatively impact plant productivity. These stresses can exacerbate deficiencies, and effective management strategies, including foliar nutrient applications and the use of protective agents, are essential to improve plant resilience under challenging conditions [38].

5. Plant-soil interaction

Plant-soil interactions are fundamental in sustaining plant growth and maintaining soil health, forming a complex and dynamic relationship where plants influence the soil environment impacting their development. These interactions are mediated through various biophysical, chemical, and biological processes that vary across spatial and temporal scales. Understanding these interactions is critical for enhancing agricultural productivity and ensuring ecosystem stability.

5.1 Plant-soil feedback (PSFs)

Plant-soil feedbacks (PSFs) describe the reciprocal interactions between plants and soils that can either enhance or inhibit plant growth and soil health (example as seen in **Figure 1**). These feedbacks involve a complex network of interactions among plant roots, soil microbes, and the surrounding soil environment.

Positive PSFs occur when plants modify the soil in ways that enhance their growth or that of similar species. For instance, leguminous plants establish symbiotic relationships with nitrogen-fixing bacteria, enriching soil nitrogen content, which benefits subsequent plantings. Mycorrhizal fungi also form mutualistic relationships with plant roots, increasing nutrient and water uptake, particularly phosphorus [40].

Negative PSFs arise when plants alter the soil environment in ways that reduce their growth or that of subsequent plants. This can occur through pathogen buildup, allelopathy (the release of harmful chemicals by plant roots), or nutrient depletion. Continuous monocropping of species like maize can lead to the accumulation of specific pathogens, such as *Fusarium oxysporum*, which can significantly reduce yield [41].

5.2 Mechanisms of soil-plant interactions

Plants release various organic compounds, including sugars, amino acids, and organic acids, into the soil through their roots. These exudates serve as a food source for soil microbes, stimulating microbial activity and enhancing nutrient cycling and

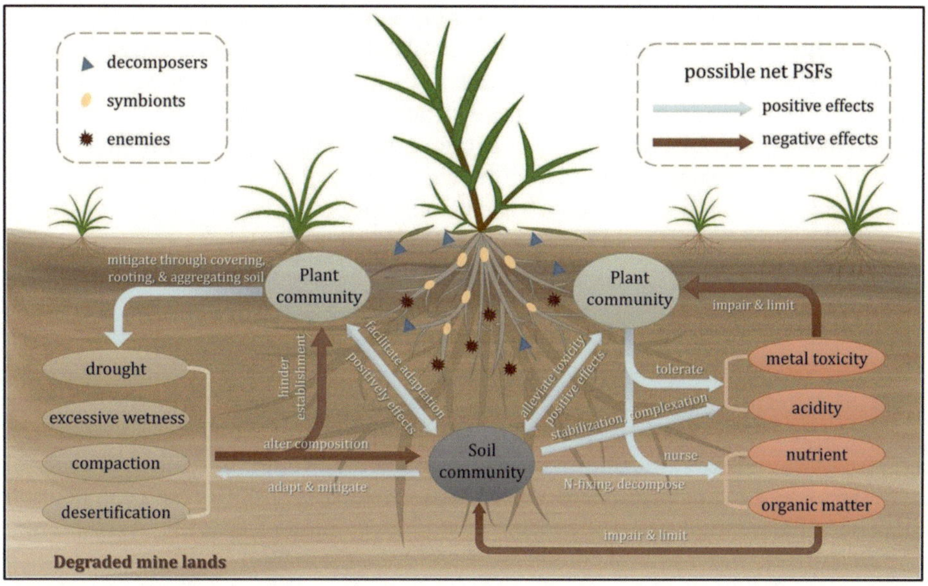

Figure 1.
Plant-soil interaction diagram [39].

availability [42]. Additionally, root exudates are crucial in soil aggregation, improving soil structure and water retention [43].

Additionally, the soil microbiome, comprising bacteria, fungi, and other microorganisms, plays a vital role in plant health. Beneficial microbes can suppress soil-borne pathogens, fix atmospheric nitrogen, and solubilize phosphorus, thereby making these nutrients more available to plants. The interactions between plant roots and these microbes are essential for maintaining soil fertility and plant productivity [44]. Soil texture, structure, and porosity directly influence plant growth. Well-structured soils with adequate porosity allow for better root penetration, water infiltration, and nutrient uptake. Conversely, compacted soils can restrict root growth and reduce the availability of essential nutrients [45].

The availability of nutrients in the soil is influenced by several factors, including pH, organic matter content, and cation exchange capacity (CEC). Soil pH, for instance, affects nutrient solubility, with most nutrients being more available in slightly acidic to neutral pH ranges. CEC, on the other hand, determines the soil's ability to retain and supply essential cations like potassium and calcium to plants [46]. These factors are crucial considerations in sustainable agriculture.

5.3 Managing plant-soil interactions for sustainable agriculture

Effective management of plant-soil interactions is key to sustainable agriculture. Practices such as crop rotation, intercropping, and the use of cover crops can balance positive and negative PSFs, promoting long-term soil health and agricultural productivity. Additionally, the application of organic amendments, such as compost and biofertilizers, can enhance soil microbial diversity and function, further supporting sustainable plant-soil interactions [47]. Advanced PSF management strategies involve manipulating soil microbial communities to favor beneficial interactions. For example, inoculating soils with mycorrhizal fungi or nitrogen-fixing bacteria can enhance nutrient availability and plant growth. Adjusting soil pH through liming or the use of acidifying agents can also optimize nutrient availability and microbial activity [48].

Plant-soil interactions significantly contribute to the resilience and stability of agricultural ecosystems. By improving soil structure, enhancing nutrient cycling, and promoting microbial diversity, these interactions help maintain soil fertility and reduce the need for chemical inputs. Moreover, healthy plant-soil interactions can mitigate environmental impacts, such as nutrient runoff and soil erosion, contributing to the long-term sustainability of agricultural systems [49].

6. Fertilization strategies for crop production

Fertilization is crucial for agricultural productivity, especially given the global need for higher crop yields and the decreasing availability of arable land. While fertilizers, particularly inorganic ones, are key to enhancing productivity, their widespread use presents significant environmental challenges, such as pollution and soil degradation. The uneven distribution of fertilizer use globally underscores the necessity for more sustainable and equitable practices. To address these issues, a shift in fertilization strategies has been advocated, focusing on a deeper understanding of nutrient roles, sources, and plant uptake processes [50].

6.1 Inorganic fertilization

Inorganic fertilization involves using synthetic fertilizers to supply plants with essential nutrients, such as nitrogen (N), phosphorus (P), and potassium (K) along with other micronutrients like sulfur, iron, and zinc. Regions like Africa face a pressing need to boost food production by 4% annually to meet the growing population's demands. This necessitates a significant increase in chemical fertilizer use, from an average of 10 to 50 kg/ha, requiring an 18% annual rise in fertilizer usage. Research indicates that Africa heavily relies on inorganic fertilizers to address nutrient losses, which can exceed 10 kg of nitrogen, 4 kg of phosphorus, and 10 kg of potassium per hectare annually in most sub-Saharan African countries. East Africa experiences the highest depletion rates, with losses surpassing 40 kg of nitrogen, 15 kg of phosphorus, and 40 kg of potassium per hectare [51]. Proper application of fertilizers, informed by soil and crop production knowledge, can help mitigate environmental impacts. Various commercial fertilizers are available, differing in nutrient content and release mechanisms, and can be categorized into nitrogen, phosphorus, potassium, sulfur, calcium, and magnesium fertilizers, some containing multiple nutrients [52].

6.1.1 Nitrogen, phosphorus, and potassium fertilizers

Nitrogen fertilizers, crucial for plant growth and leaf development, are widely used in Africa. Common sources include ammonium sulfate, ammonium phosphates, and urea. Urea, as shown in **Figure 2**, contains 45–46% nitrogen and, is particularly valued for its cost-effectiveness and ability to convert to ammonium in the soil, reducing leaching risks. Phosphorus fertilizers, such as rock phosphates and superphosphates (SSP and TSP), are vital for plant root development and energy transfer. Rock phosphates are more effective in acidic soils, while superphosphates provide readily available phosphorus, essential for various crops. Potassium fertilizers, such as potassium chloride (KCl) and potassium sulfate (K_2SO_4), support plant processes like

Figure 2.
An example of commonly used inorganic fertilizers. (a) Sodium nitrate. (b) Urea. (c) Muriate of potash.

water regulation and enzyme activation. Potassium chloride is the most widely used potassium fertilizer due to its high solubility and effectiveness.

6.1.2 Benefits and potential environmental risks of inorganic fertilizers

While inorganic fertilizers are crucial for global food production, their overuse poses significant environmental risks. Intensive application can lead to air, water, and soil pollution, land degradation, and increased greenhouse gas emissions. Long-term use affects soil microbiology, altering pH and impacting bacterial diversity and soil microbial activity. Moreover, excessive nitrogen fertilization can lead to nutrient leaching, contributing to water eutrophication, soil acidity, and reduced crop yields [53]. Therefore, while inorganic fertilizers are key in boosting agricultural productivity, their use must be carefully managed to avoid adverse environmental consequences. Combining them with organic fertilization and adopting practices such as INM can enhance nutrient use efficiency and promote sustainable farming systems.

6.2 Organic fertilization

Organic fertilizers are derived from animal wastes, plant residues, treated sewage sludge, and leguminous plants. Organic amendments improve soil nutrient availability, organic matter content, and microbial activity, thereby enhancing crop yield and growth [54]. About 2.4 million farmers globally practice organic agriculture on 50.9 million hectares, representing 1.1% of total agricultural land [55]. The choice of organic fertilizers depends on crop type, season, soil chemistry, and farmers' experiences. Research has explored the agronomic potential of various organic fertilizers on crops and animal-based organic fertilizers [56]. Organic fertilizers support a diverse soil ecosystem, providing essential nutrients for beneficial microorganisms that maintain healthy soil and suppress harmful pathogens. They enhance soil fertility by improving porosity and water-holding capacity while promoting microbial activity. The slow release of nutrients from organic materials ensures a steady supply of essential elements, preventing nutrient depletion and promoting balanced nutrient levels in the soil [57].

6.3 Integrated nutrient management

Integrated nutrient management practices have gained significant attention in agriculture due to their potential to enhance crop productivity and sustainability. By combining different nutrient sources, farmers can improve soil health, increase nutrient availability to crops, and reduce the reliance on synthetic fertilizers. This improves the economic viability of farming and contributes to environmental conservation.

Research has demonstrated that INM can lead to better nutrient utilization by crops, improved soil structure and fertility, reduced greenhouse gas emissions from synthetic fertilizer use, and decreased risk of nutrient runoff into water bodies [58, 59]. However, the successful implementation of INM requires careful consideration of various factors such as soil type, crop type, climate conditions, availability of inputs, and farmer knowledge. Adequate training and support are essential for farmers to adopt INM practices effectively.

6.4 Best practices for maximizing nutrient use efficiency and minimizing environmental impact

Combining organic and inorganic fertilizers can optimize soil fertility and crop productivity. Organic fertilizers, like manure and compost, enhance soil structure and provide a long-term nutrient supply, while inorganic fertilizers offer a rapid source of essential nutrients. Integrated Nutrient Management (INM) effectively combines these benefits, ensuring sustained crop nutrition and addressing immediate nutrient deficiencies. INM not only provides a sustainable and economical approach to nutrient management but also helps reduce reliance on chemical fertilizer [60]. However, while the individual effects of INM on crop productivity and soil quality are well-documented, its overall impact on factors such as soil bulk density and microbial biomass carbon is still not fully understood. Proper fertilizer management, including balanced fertilization based on soil testing, is crucial for maximizing nutrient use efficiency and minimizing environmental impacts like greenhouse gas emissions and nutrient runoff. Precision agriculture technologies and practices like cover cropping and crop rotation further enhance nutrient management, reducing waste and protecting aquatic environments; application of fertilizer in croplands is a major source of anthropogenic greenhouse gas emissions [61].

7. Innovative practices for sustainable intensification

7.1 Enhanced crop varieties

Recent advancements in crop breeding focus on developing varieties that utilize nutrients more efficiently and are resilient to climate stresses, ensuring stable yields [62]. For example, the development of drought-resistant maize in sub-Saharan Africa has significantly increased food security in the region [63]. These advancements in crop breeding have the potential to transform agriculture, enabling farmers to produce more with fewer resources while minimizing environmental impact. As research advances, we can expect to see even more innovative solutions that address the pressing challenges of food security and sustainability. In addition, biotechnology, in recent years, has also facilitated the creation of crops with precise genome modifications, allowing for targeted improvements in yield, disease resistance, and nutritional quality. Notable examples include Golden Rice, which is enriched with beta-carotene to combat vitamin A deficiency [64].

7.2 Agroecological practices

Diversifying crop species helps buffer against market and climatic risks, promoting a more resilient agricultural system. Agroforestry, which integrates trees and shrubs with crops and livestock, offers numerous benefits including improved soil fertility, enhanced biodiversity, and better microclimates. Trees in agroforestry systems can act as windbreaks, reduce soil erosion, and enhance water infiltration. Additionally, they can provide shade for crops and livestock, creating a more favorable microclimate. Intercropping, or growing multiple crops together, maximizes resource use and reduces pest pressures [65]. For example, integrating nitrogen-fixing trees with staple crops in Zambia has significantly improved yields and soil health. Such systems also provide additional products like fruits, fodder, and timber, which can diversify income sources for farmers and improve their livelihoods [66]. Moreover, these practices contribute to carbon sequestration, aiding in climate change mitigation. Agroforestry systems, for instance, have a higher capacity for carbon storage than traditional farming systems.

7.3 Conservation tillage

Conservation tillage techniques, such as no-till and reduced-till farming, are revolutionizing agricultural practices by maintaining soil structure, enhancing organic matter content, and significantly reducing erosion. For instance, in the United States, conservation tillage practices have reduced soil erosion by up to 90% in some areas [67]. The benefits of conservation tillage extend beyond soil preservation. By minimizing soil disturbance, these techniques help maintain the habitat for beneficial soil organisms, which play a critical role in nutrient cycling and soil health. This approach conserves soil moisture, which is crucial during dry periods and reduces farmers' fuel and labor costs, making it an economically viable option. Moreover, conservation tillage is pivotal in mitigating climate change. Reducing the need for frequent plowing, lowers greenhouse gas emissions from farm machinery. Additionally, the increased organic matter in the soil acts as a carbon sink, sequestering carbon dioxide from the atmosphere. This dual benefit of enhancing soil health while combating climate change makes conservation tillage a cornerstone of sustainable agriculture. In practice, farmers adopting conservation tillage have reported improved crop resilience to drought and extreme weather events [68]. This resilience is particularly vital in the face of global climate change, ensuring food security and sustainable farming livelihoods.

7.4 Technological advances

Integration artificial intelligence (AI), big data analytics, and remote sensing technologies has revolutionized the monitoring and management of agroecosystems. These technologies enable real-time data collection and analysis, which helps predict crop performance, identify pest and disease outbreaks, and optimize irrigation and nutrient applications. Such innovations lead to more sustainable and efficient farming practices. For instance, in California's Central Valley, remote sensing data from satellites is used to monitor crop health and water usage, optimize irrigation practices and conserve water. AI has enabled the development of predictive models that analyze vast datasets to forecast agricultural trends, improve crop management decisions, and enhance resilience to climatic variability. Practical tools include AI-driven platforms like Climate FieldView, which provides farmers with detailed insights into field conditions, and Plantix, an app that uses AI to diagnose plant diseases from photos taken by farmers, offering treatment recommendations [69, 70]. These digital tools support sustainable agricultural practices by minimizing resource wastage and environmental degradation. As these technologies advance, they promise to transform agriculture into a more efficient, sustainable, and resilient industry.

8. Enhancing policy and institutional frameworks for sustainable agriculture in Africa

Enhancing the policy and institutional frameworks for sustainable agriculture for example in Africa is crucial given the continent's heavy reliance on agriculture for economic stability, food security, and environmental sustainability. Robust frameworks are necessary to provide the guidelines, incentives, and regulations to effectively promote sustainable agricultural practices. These frameworks are instrumental in integrating sustainable agriculture into national strategies by ensuring both vertical alignment with international directives and horizontal integration across various governmental sectors.

8.1 Vertical and horizontal policy coherence

In the African context, aligning international and local policies is critical. International agreements, such as the Paris Agreement and the Sustainable Development Goals (SDGs), serve as overarching frameworks guiding national policies. For African nations, this involves adapting global standards to local conditions, which vary widely in terms of ecology, economy, and society [71]. Horizontal integration, on the other hand, ensures that agricultural policies are in harmony with policies from other sectors like water management, economic development, and environmental protection. This coherence is essential for ensuring sustainable agricultural policies contribute to broader development goals without compromising environmental sustainability or social equity [72].

8.2 Global policies and agreements

Global organizations, including the Food and Agriculture Organization (FAO) and the Consultative Group on International Agricultural Research (CGIAR), play pivotal roles in global agricultural policies. They set standards and provide technical support to facilitate the implementation of sustainable practices across Africa. These organizations ensure that policies are scientifically sound and practically viable [73]. Agreements like the Paris Agreement and the SDGs significantly influence national policies across Africa. These global frameworks drive efforts to reduce agricultural emissions, enhance carbon sequestration, and promote sustainable land and soil management practices that can adapt to the continent's climatic challenges [74].

8.3 National and local policies

At the national level, policies such as Kenya's Climate Smart Agriculture Strategy and Ethiopia's Green Economy Strategy provide frameworks to support sustainable agricultural practices through financial and technical assistance. These strategies are akin to similar programs found in countries like the United States and Germany [75]. Local governance and community involvement are key to effectively implementing sustainable practices. Policies must be grounded in local realities to ensure their effectiveness and acceptance. Community-led initiatives, such as Tanzania's community forest management, highlight the potential of local involvement in enhancing ecosystems and providing sustainable livelihoods [76]. Africa faces significant challenges in coordinating policies across different government levels and sectors, often resulting in conflicts, delays, and inefficiencies. Resistance to change and limited adaptation of policies to local conditions further hinder their effectiveness. Frequently, policies do not consider the unique ecological, social, and economic conditions of local areas, rendering them less effective or even irrelevant [77].

8.4 Strategies for effective policy frameworks

Developing policies through inclusive processes involving all stakeholders is essential. These policies must be flexible, adaptive, and capable of evolving with new scientific findings and changing circumstances. Regular scientific reviews and the establishment of bridging institutions can enhance the integration of scientific research into policymaking, ensuring that policies are up-to-date with the latest advancements in agricultural science [78]. The future of sustainable agricultural policies in Africa will likely involve increased integration of digital tools in policy implementation. Technologies like AI, big data, and remote sensing are being employed to improve the precision and efficiency of policy enforcement and monitoring across the continent. As global challenges such as climate change intensify, regulatory frameworks are expected to evolve to provide greater support for green technologies and align more closely with global climate agendas, ensuring a unified response to these challenges [79].

9. Future directions and research needs

Several key research areas and future directions are crucial to advance sustainable intensification. First, expanding the adoption of precision agriculture, particularly in regions reliant on traditional practices, requires focusing on affordable, scalable technologies for smallholder farmers. Integrating AI, machine learning, and big data into agriculture can enhance predictions for crop health and resource use [80]. With climate change intensifying, research into climate resilience is critical. Developing drought, heat, and flood-resistant crops and optimizing nutrient management under varying climates are priorities. Soil health and biodiversity also need deeper investigation, focusing on the interactions between soil microorganisms, plant roots, and sustainable practices like conservation agriculture and agroforestry [81].

Refining Integrated Nutrient Management (INM) strategies for different agro-ecological zones and exploring biofertilizers are essential for sustainable nutrient management [82]. Additionally, research should explore supportive policies and community-led initiatives that integrate traditional knowledge with modern practices, ensuring that strategies are locally appropriate and widely adopted. These initiatives will help ensure that policies and practices are well-suited to local conditions and widely adopted. Finally, understanding the economic viability and social impacts of sustainable intensification, including labor dynamics and gender roles, is vital for designing economically and socially sustainable interventions [83].

10. Conclusions

Sustainable intensification represents a vital paradigm shift in agriculture, aiming to balance the increasing global food demand with the need to preserve environmental health. This approach is not just about increasing productivity, but about doing so in a way that maintains and enhances the ecological systems upon which agriculture depends. Key principles such as diversification, precision agriculture, and integrated nutrient management are central to this approach. These practices, when effectively implemented, can lead to enhanced productivity with minimal environmental impact, improved soil health, and increased resilience to climate change. The integration of agroecological practices, such as agroforestry and conservation agriculture, further supports the sustainability of agricultural systems by promoting biodiversity and ecological balance. However, the success of sustainable intensification relies heavily on supportive policies, community engagement, and ongoing research and innovation. Policies that incentivize sustainable practices and participatory approaches involving local communities, are essential for the widespread adoption of sustainable intensification.

Additionally, research focusing on refining sustainable practices, understanding the socio-economic impacts, and developing new technologies will be crucial in addressing the challenges of global food security and environmental sustainability. In conclusion, sustainable intensification offers a promising pathway to meet the dual challenges of feeding a growing global population and preserving the planet's ecological balance. Integrating productivity and environmental health provides a framework for resilient and sustainable agriculture in the face of climate change and other global challenges. Continued research, innovation, and collaboration across sectors will be key to realizing the full potential of a sustainable and resilient future.

Acknowledgements

The authors gratefully acknowledge the Climate and Clean Air Coalition (CCAC) of the United Nations Environment Program for their generous funding and indispensable support, which were pivotal to the success of this research. The authors acknowledge the use of Grammarly, ChatGPT, and Perplexity AI for language polishing and reference support in the manuscript.

Conflict of interest

The author declares no conflict of interest.

Author details

Kwaku Onwona-Hwesofour Asante, Jeannette Aduhene-Chinbuah,
Clement Oppong Peprah, Kennedy Agyeman, Felix Frimpong, Eric Owusu Danquah and Ralph Bam*
Council for Scientific and Industrial Research-Crops Research Institute, Kumasi, Ghana

*Address all correspondence to: ralphbam@yahoo.com

References

[1] Foley JA et al. Solutions for a cultivated planet. Nature. 2011;**478**(7369):337-342. DOI: 10.1038/nature10452

[2] Tscharntke T, Klein AM, Kruess A, Steffan-Dewenter I, Thies C. Landscape perspectives on agricultural intensification and biodiversity-ecosystem service management. Ecology Letters. 2005;**8**(8):857-874. DOI: 10.1111/j.1461-0248.2005.00782.x

[3] Tilman D, Balzer C, Hill J, Befort BL. Global food demand and the sustainable intensification of agriculture. Proceedings of the National Academy of Sciences. 2011;**108**(50):20260-20264. DOI: 10.1073/pnas.1116437108

[4] Campbell BM, Thornton P, Zougmoré R, van Asten P, Lipper L. Sustainable intensification: What is its role in climate smart agriculture? Current Opinion in Environment Sustainability. 2014;**8**:39-43. DOI: 10.1016/j.cosust.2014.07.002

[5] Pretty J et al. Global assessment of agricultural system redesign for sustainable intensification. Nature Sustainability. 2018;**1**(8):441-446. DOI: 10.1038/s41893-018-0114-0

[6] Pretty J, Toulmin C, Williams S. Sustainable intensification in African agriculture. International Journal of Agricultural Sustainability. 2011;**9**(1):5-24. DOI: 10.3763/ijas.2010.0583

[7] Veste M et al. The need for sustainable agricultural land-use systems: Benefits from integrated agroforestry systems. In: von Maltitz GP, et al., editor. Sustainability of Southern African Ecosystems under Global Change. Ecological Studies. Vol. 248. Cham: Springer; 2024. pp. 587-623. DOI: 10.1007/978-3-031-10948-5_21

[8] McBratney A, Whelan B, Ancev T, Bouma J. Future directions of precision agriculture. Precision Agriculture. 2005;**6**(1):7-23. DOI: 10.1007/S11119-005-0681-8/METRICS

[9] Evenson RE, Gollin D. Assessing the impact of the green revolution, 1960 to 2000. Science (1979). 2003;**300**(5620):758-762. DOI: 10.1126/science.1078710

[10] Pinstrup-Andersen P, Hazell PBR. The impact of the green revolution and prospects for the future. Food Reviews International. 1985;**1**(1):1-25. DOI: 10.1080/87559128509540765

[11] Hou D, Bolan NS, Tsang DCW, Kirkham MB, O'Connor D. Sustainable soil use and management: An interdisciplinary and systematic approach. Science of the Total Environment. 2020;**729**:138961. DOI: 10.1016/j.scitotenv.2020.138961

[12] Delgado JA, Short NM, Roberts DP, Vandenberg B. Big data analysis for sustainable agriculture on a geospatial cloud framework. Frontiers in Sustainable Food Systems. Jul 2019;**3**:54. DOI: 10.3389/fsufs.2019.00054

[13] Krishna KR. Agricultural Drones. Waretown, NJ: Apple Academic Press; 2018. DOI: 10.1201/9781315195520

[14] Lowenberg-DeBoer J, Erickson B. Setting the record straight on precision agriculture adoption. Agronomy Journal. 2019;**111**(4):1552-1569. DOI: 10.2134/agronj2018.12.0779

[15] Erlangga E, Machuku O, Jun Dahino C. A review article on the impact and challenges of mobile phone usage on agricultural production in

Africa. Cogent Food & Agriculture. Dec 2023;**9**(2):2273634. DOI: 10.1080/23311932.2023.2273634

[16] Cisternas I, Velásquez I, Caro A, Rodríguez A. Systematic literature review of implementations of precision agriculture. Computers and Electronics in Agriculture. 2020;**176**:105626. DOI: 10.1016/j.compag.2020.105626

[17] Philippine Economic Update, December 2012: Accelerating Reforms to Sustain Growth. Manila. Available from: https://hdl.handle.net/10986/26742: World Bank; 2012 [Accessed: July 15, 2024]

[18] Raj R, Kar S, Nandan R, Jagarlapudi A. Precision agriculture and unmanned aerial vehicles (UAVs). In: Unmanned Aerial Vehicle: Applications in Agriculture and Environment. Cham: Springer International Publishing; 2020. pp. 7-23. DOI: 10.1007/978-3-030-27157-2_2

[19] Lampach N, To-The N, Nguyen-Anh T. Technical efficiency and the adoption of multiple agricultural technologies in the mountainous areas of northern Vietnam. Land Use Policy. 2021;**103**:105289. DOI: 10.1016/j.landusepol.2021.105289

[20] Surendran U, Nagakumar KCV, Samuel MP. Remote sensing in precision agriculture. In: Digital Agriculture. Cham: Springer International Publishing; 2024. pp. 201-223. DOI: 10.1007/978-3-031-43548-5_7

[21] Nakai J. Food and Agriculture Organization of the United Nations and the sustainable development goals. Sustainable Development (Rome). 2018;**22**:1-450

[22] Souza Piao R, Silva VL, Navarro del Aguila I, de Burgos Jiménez J. Green growth and agriculture in Brazil. Sustainability. 2021;**13**(3):1162. DOI: 10.3390/su13031162

[23] Van den Berg H, Jiggins J. Investing in farmers—The impacts of farmer field schools in relation to integrated pest management. World Development. 2007;**35**(4):663-686. DOI: 10.1016/j.worlddev.2006.05.004

[24] Mäder P, Fließbach A, Dubois D, Gunst L, Fried P, Niggli U. Soil fertility and biodiversity in organic farming. Science (1979). 2002;**296**(5573):1694-1697. DOI: 10.1126/SCIENCE.1071148

[25] Francis C et al. Agroecology: The ecology of food systems. Journal of Sustainable Agriculture. Jul 2003;**22**(3):99-118. DOI: 10.1300/J064v22n03_10

[26] das Nair R, Landani N. Making Agricultural Value Chains more Inclusive Through Technology and Innovation. Finland: UNU-WIDER; 2020. DOI: 10.35188/UNU-WIDER/2020/795-8

[27] Snapp SS et al. Participatory action research generates knowledge for Sustainable Development Goals. Frontiers in Ecology and the Environment. 2023;**21**(7):341-349. DOI: 10.1002/fee.2591

[28] Shawki N, Schnyder M. Understanding polycentrically organized movements: An analysis of the Slow Food and Fair Trade movements in Europe. Journal of Civil Society. 2023;**19**(2):156-174. DOI: 10.1080/17448689.2023.2206153

[29] Marten GG. Productivity, stability, sustainability, equitability and autonomy as properties for agroecosystem assessment. Agricultural Systems. 1988;**26**(4):291-316. DOI: 10.1016/0308-521X(88)90046-7

[30] Belete T, Yadete E. Effect of mono cropping on soil health and fertility management for sustainable agriculture practices: A review. Journal of Plant Sciences. 2023;**11**:192-197

[31] Shah KK et al. Diversified crop rotation: An approach for sustainable agriculture production. Advances in Agriculture. 2021;**2021**:1-9. DOI: 10.1155/2021/8924087

[32] Chain-Guadarrama A, Martínez-Salinas A, Aristizábal N, Ricketts TH. Ecosystem services by birds and bees to coffee in a changing climate: A review of coffee berry borer control and pollination. Agriculture, Ecosystems and Environment. 2019;**280**:53-67. DOI: 10.1016/j.agee.2019.04.011

[33] Khan ZR et al. Achieving food security for one million sub-Saharan African poor through push–pull innovation by 2020. Philosophical Transactions of the Royal Society, B: Biological Sciences. 2014;**369**(1639):20120284. DOI: 10.1098/rstb.2012.0284

[34] Shiferaw BA, Okello J, Reddy RV. Adoption and adaptation of natural resource management innovations in smallholder agriculture: Reflections on key lessons and best practices. Environment, Development and Sustainability. 2009;**11**(3):601-619. DOI: 10.1007/s10668-007-9132-1

[35] Lazali M et al. CROSYMED project: Enhancing nutrient use efficiency through legumes in agroecosystems of the Mediterranean Basin. Sustainability. 2021;**13**(9):4695. DOI: 10.3390/su13094695

[36] Rumpel C et al. The 4p1000 initiative: Opportunities, limitations and challenges for implementing soil organic carbon sequestration as a sustainable development strategy. Ambio. 2020;**49**(1):350-360. DOI: 10.1007/s13280-019-01165-2

[37] Andrews SS, Karlen DL, Cambardella CA. The soil management assessment framework. Soil Science Society of America Journal. 2004;**68**(6):1945-1962. DOI: 10.2136/SSSAJ2004.1945

[38] Rusan MJ. Soil and nutrient management under saline conditions. In: Biosaline Agriculture as a Climate Change Adaptation for Food Security. Cham: Springer; 2023. pp. 71-85. DOI: 10.1007/978-3-031-24279-3_4

[39] Zhu S-C et al. Plant-soil feedbacks for the restoration of degraded mine lands: A review. Frontiers in Microbiology. Jan 2022;**12**:751794. DOI: 10.3389/fmicb.2021.751794

[40] Koudahe K, Allen SC, Djaman K. Critical review of the impact of cover crops on soil properties. International Soil and Water Conservation Research. 2022;**10**(3):343-354. DOI: 10.1016/j.iswcr.2022.03.003

[41] Bennett JA, Klironomos J. Mechanisms of plant–soil feedback: Interactions among biotic and abiotic drivers. New Phytologist. 2019;**222**(1): 91-96. DOI: 10.1111/nph.15603

[42] Dakora FD, Phillips DA. Root exudates as mediators of mineral acquisition in low-nutrient environments. In: Food Security in Nutrient-Stressed Environments: Exploiting Plants' Genetic Capabilities. Developments in Plant and Soil Sciences. Vol. 95. Dordrecht: Springer; 2002. pp. 201-213. DOI: 10.1007/978-94-017-1570-6_23

[43] Jones DL, Nguyen C, Finlay RD. Carbon flow in the rhizosphere: Carbon trading at the soil-root interface.

Plant and Soil. 2009;**321**(1-2):5-33. DOI: 10.1007/S11104-009-9925-0/METRICS

[44] Parnell JJ et al. From the lab to the farm: An industrial perspective of plant beneficial microorganisms. Frontiers in Plant Science. 2016;7(AUG2016):209596. DOI: 10.3389/FPLS.2016.01110/BIBTEX

[45] Brady N, Weil R. Elements of the Nature and Properties of Soils [Online]. 2004. Available from: http://faculty.washington.edu/zabow/ESC210/210Syllabus_2007.doc [Accessed: August 21, 2024]

[46] Tisdale S, Nelson W, Beaton J. Soil Fertility and Fertilizers [Online]. 1985. Available from: https://www.cabidigitallibrary.org/doi/full/10.5555/19851998321 [Accessed: August 21, 2024]

[47] Li Y, Tian D, Feng G, Yang W, Feng L. Climate change and cover crop effects on water use efficiency of a corn-soybean rotation system. Agricultural Water Management. 2021;**255**:107042. DOI: 10.1016/j.agwat.2021.107042

[48] Page V, Feller U. Heavy metals in crop plants: Transport and redistribution processes on the whole plant level. Agronomy. 2015;**5**(3):447-463. DOI: 10.3390/agronomy5030447

[49] Lal R. Soil carbon sequestration impacts on global climate change and food security. Science (1979). 2004;**304**(5677):1623-1627. DOI: 10.1126/SCIENCE.1097396/SUPPL_FILE/LAL.SOM.PDF

[50] Tilman D, Clark M. Food, agriculture & the environment: Can we feed the world & save the earth? Daedalus. 2015;**144**(4):8-23. DOI: 10.1162/DAED_a_00350

[51] Bindraban PS, Dimkpa C, Nagarajan L, Roy A, Rabbinge R. Revisiting fertilisers and fertilisation strategies for improved nutrient uptake by plants. Biology and Fertility of Soils. 2015;**51**(8):897-911. DOI: 10.1007/s00374-015-1039-7

[52] Amalu UC. Food security: Sustainable food production in sub-Saharan Africa. Outlook on Agriculture. 2002;**31**(3):177-185. DOI: 10.5367/000000002101294029

[53] Færge J, Magid J. Evaluating NUTMON nutrient balancing in sub-Saharan Africa. Nutrient Cycling in Agroecosystems. 2004;**69**(2):101-110. DOI: 10.1023/B:FRES.0000029680.97610.51

[54] Kumar R, Prakash OG. The Impact of Chemical Fertilizers on Our Environment and Ecosystem [Online]. Available from: https://www.researchgate.net/publication/331132826 [Accessed: August 07, 2024]

[55] Ren N, Wang Y, Ye Y, Zhao Y, Huang Y, Fu W, et al. Effects of continuous nitrogen fertilizer application on the diversity and composition of rhizosphere soil Bacteria. Frontiers in Microbiology. 2020;**11**:1948

[56] Jote CA. The impacts of using inorganic chemical fertilizers on the environment and human health. Organic and Medicinal Chemistry International Journal. 2023;**13**(3):555864

[57] García-Orenes F, Roldán A, Morugán-Coronado A, Linares C, Cerdà A, Caravaca F. Organic fertilization in traditional Mediterranean grapevine orchards mediates changes in soil microbial community structure and enhances soil fertility. Land Degradation and Development. 2016;**27**(6):1622-1628. DOI: 10.1002/ldr.2496

[58] The World of Organic Agriculture 2024. FAO [Online]. Available from: https://www.fao.org/family-farming/detail/en/c/1678159/ [Accessed: August 07, 2024]

[59] Magid J, Pedersen KE, Hansen M, Cedergreen N, Brandt KK. Comparative assessment of the risks associated with use of manure and sewage sludge in Danish agriculture. Advances in Agronomy. 2020;**164**:289-334

[60] Campbell CA, Biederbeck VO, Selles F, Schnitzer M, Stewart JWB. Effect of manure and P fertilizer on properties of a black Chernozem in Southern Saskatchewan. Canadian Journal of Soil Science. 1986;**66**(4):601-614. DOI: 10.4141/cjss86-060

[61] Paramesh V, Mohan Kumar R, Rajanna GA, Gowda S, Nath AJ, Madival Y, et al. Integrated nutrient management for improving crop yields, soil properties, and reducing greenhouse gas emissions. Frontiers in Sustainable Food Systems. 2023;7:1173258

[62] Rahman MM et al. Integrated nutrient management enhances productivity and nitrogen use efficiency of crops in acidic and Charland soils. Plants. 2021;**10**(11):2547. DOI: 10.3390/plants10112547

[63] Chaubey AK, Mishra S, Singh SK, Chaubey C, Pandey KP. Integrated nutrients management for future production: A review. International Journal of Environment and Climate Change. 2023;**13**(11):1770-1779. DOI: 10.9734/ijecc/2023/v13i113333

[64] Selim MM. Introduction to the integrated nutrient management strategies and their contribution to yield and soil properties. International Journal of Agronomy. 2020;**2020**:1-14. DOI: 10.1155/2020/2821678

[65] Tester M, Langridge P. Breeding technologies to increase crop production in a changing world. Science (1979). 2010;**327**(5967):818-822. DOI: 10.1126/science.1183700

[66] Gebre GG, Rahut DB, Aryal JP, Mawia H. Potential impact of scaling adaptation strategies for drought stress: A case of drought-tolerant maize varieties in Tanzania. International Journal of Agricultural Sustainability. 2023;**21**(1):2189396

[67] Bouis HE, Saltzman A. Improving nutrition through biofortification: A review of evidence from HarvestPlus, 2003 through 2016. Global Food Security. 2017;**12**:49-58. DOI: 10.1016/j.gfs.2017.01.009

[68] Brookes G, Barfoot P. Environmental impacts of genetically modified (GM) crop use 1996-2016: Impacts on pesticide use and carbon emissions. GM Crops & Food. 2018;**9**(3):109-139. DOI: 10.1080/21645698.2018.1476792

[69] Huss CP, Holmes KD, Blubaugh CK. Benefits and risks of intercropping for crop resilience and pest management. Journal of Economic Entomology. 2022;**115**(5):1350-1362. DOI: 10.1093/jee/toac045

[70] Garrity DP et al. Evergreen agriculture: A robust approach to sustainable food security in Africa. Food Security. 2010;**2**(3):197-214. DOI: 10.1007/s12571-010-0070-7

[71] Kumar V, Mahajan G, Dahiya S, Chauhan BS. Challenges and opportunities for weed management in no-till farming systems. In: No-Till Farming Systems for Sustainable Agriculture. Cham: Springer International Publishing; 2020. pp. 107-125. DOI: 10.1007/978-3-030-46409-7_7

[72] Shrestha JIBAN, Subedi SUBASH, Timsina KP, Chaudhary A, Kandel M, Tripathi S. Conservation agriculture as an approach towards sustainable crop production: A review. Farming and Management. 2020;**5**(1):7-15

[73] Hampf AC, Nendel C, Strey S, Strey R. Biotic yield losses in the Southern Amazon, Brazil: Making use of smartphone-assisted plant disease diagnosis data. Frontiers in Plant Science. 2021;**12**:621168

[74] Tiefenbacher A, Sandén T, Haslmayr HP, Miloczki J, Wenzel W, Spiegel H. Optimizing carbon sequestration in croplands: A synthesis. Agronomy. 2021;**11**(5):882

[75] Vargas VR, Lawthom R, Prowse A, Randles S, Tzoulas K. Implications of vertical policy integration for sustainable development implementation in higher education institutions. Journal of Cleaner Production. 2019;**235**:733-740. DOI: 10.1016/j.jclepro.2019.07.022

[76] Pawlak K, Kołodziejczak M, Xie Y. Horizontal integration in the agricultural sector as a factor increasing its competitiveness—Experience from Poland. Eastern European Countryside. 2019;**25**(1):195-232. DOI: 10.12775/eec.2019.008

[77] Pingali P, Kelley T. The role of international agricultural research in contributing to global food security and poverty alleviation: The case of the CGIAR. In: Handbook of Agricultural Economics. Vol. 3. Elsevier; 2007. pp. 2381-2418

[78] Nhemachena C, Matchaya G, Nhemachena C, Karuaihe S, Muchara B, Nhlengethwa S. Measuring baseline agriculture-related sustainable development goals index for southern Africa. Sustainability. 2018;**10**(3):849. DOI: 10.3390/su10030849

[79] Ogemah VK. Sustainable agriculture: Developing a common understanding for modernization of agriculture in Africa. African Journal of Food, Agriculture, Nutrition and Development. 2017;**17**(1):11673-11690. DOI: 10.18697/AJFAND.77.16560

[80] Esteves AM, Genus A, Henfrey T, Penha-Lopes G, East M. Sustainable entrepreneurship and the sustainable development goals: Community-led initiatives, the social solidarity economy and commons ecologies. Business Strategy and the Environment. 2021;**30**(3):1423-1435. DOI: 10.1002/bse.2706

[81] Jastrzębska M, Kostrzewska M, Saeid A. Sustainable agriculture: A challenge for the future. In: Smart Agrochemicals for Sustainable Agriculture. Netherland: Elsevier; 2022. pp. 29-56. DOI: 10.1016/B978-0-12-817036-6.00002-9

[82] Snyder KA, Miththapala S, Sommer R, Braslow J. The yield gap: Closing the gap by widening the approach. Experimental Agriculture. 2017;**53**(3):445-459. DOI: 10.1017/S0014479716000508

[83] Ehlers M-H, Huber R, Finger R. Agricultural policy in the era of digitalisation. Food Policy. 2021;**100**:102019. DOI: 10.1016/j.foodpol.2020.102019

Section 4

Technology

Chapter 9

A Rhizobox-Study Elucidating Biogas-Digestate Fertilization and Soil Compaction Effects on Juvenile Maize Growth and Rhizosphere pH

Daniela Piaz Barbosa Leal, Otávio dos Anjos Leal, Stephan Blossfeld and Nicolai David Jablonowski

Abstract

Biogas digestate (BD) contains nitrogen (N), phosphorus (P), and potassium (K) and is easily and largely available in Germany and other countries in Europe. Nevertheless, few studies compare BD to mineral NPK fertilizer, particularly under soil compaction. The characteristics of these fertilizers and soil compaction may affect rhizosphere pH and root development, thereby affecting nutrient uptake by plants. This 18-day rhizobox study evaluated initial maize growth and root architecture responses to BD (derived from maize silage+chicken manure), mineral NPK, and BD + NPK fertilization under compacted (0–25 cm compacted; 25–55 cm loose) and loose (0–55 cm) substrates. Treatments promoted similar shoot biomass, specific leaf area, and sufficient shoot N, P, and K nutrition. Shoot P content in BD + NPK and NPK was 29–33% higher compared to BD. Root P content in NPK was 26% higher than in BD, independent of compaction, likely favoring root proliferation and explaining the higher root:shoot ratio in NPK. In loose soil, the main root length in NPK was 49% higher compared to BD, but rooting was deeper in BD. Rhizosphere alkalinization measured non-invasively suggested preferential maize N absorption as nitrate. Combining BD with fast-soluble P sources may provide maize performance comparable to mineral NPK.

Keywords: mineral fertilizer, rhizosphere alkalinization, non-invasive rhizosphere pH, root architecture, shoot nutrition

1. Introduction

Plants face spatial and temporal variability of biological, chemical, and physical soil properties and show the ability to adjust to such heterogeneity through their morphological and physiological plasticity [1, 2]. Typical responses of a plant subjected to soil compaction may include restriction of downward penetration of the main root axes, decrease of root elongation rate and total root length [3], increase in shoot-to-root biomass ratio [4], and, lastly, reduction of crop yield [5]. Furthermore, soil compaction affects plant water use and nutrient uptake [6–8]. Interestingly, studies reported that plants differ in the extent of compensatory root growth responses to heterogeneous soil strength [9–12].

The arable soil layer (0–25 cm) is also the main rooting zone, and its compaction has been reported as one of the most limiting plant performance factors [13–15]. Measurements such as soil bulk density and soil penetration resistance are widely used to determine soil compactness [7, 16]. It should be mentioned that critical limits for soil bulk density and soil penetration resistance vary depending mainly on soil texture, soil water content, and plant species [17]. Organic substrates, commonly used as a growth medium in greenhouse experiments, normally present low bulk density [11]. Moreover, organic substrates under compaction usually present penetration resistance values lower than the critical thresholds determined for soils, but the effects of compaction on root and shoot are presumably reproducible in organic substrate [18, 19]. In addition, a slight degree of soil compaction may be favorable for the adhesion of the soil to roots and subsequent root exploration for water and nutrients [20, 21].

Changes in rhizosphere pH are considered a critical parameter that influences the bioavailability of nutrients and toxic elements for plant uptake [22]. Roots can induce an increase or decrease of rhizosphere pH by OH^- or H^+ release, respectively, changing the equilibrium between cations and anions at the root-soil interface [23]. In general, when more cations than anions are absorbed by the roots, plants release proportionally more protons in the rhizosphere, thereby acidifying the rhizosphere pH [23]. On the other side, when more anions than cations are absorbed, either H^+ is taken up, or OH^- or HCO_3^- is excreted, hence leading to the alkalinization of the rhizosphere [22, 23]. Alkalinization or acidification of the rhizosphere and its intensity depend mainly on plant species and soil factors, such as initial bulk soil pH, soil water content, and fertilization. Nitrogen (N) uptake as nitrate (NO_3^-) or ammonium (NH_4^+) generally leads to rhizosphere alkalinization and acidification, respectively, in non-legume species such as maize (*Zea mays* L.) [24]. In this way, monitoring the rhizosphere pH allows us to improve our understanding of how plants use the nutrients released from mineral or organic fertilizers in the soil.

The elucidation of root phenotypic traits and biogeochemical processes is strongly dependent on the availability of non-invasive analytical methods [25]. In this sense, plant experiments performed in transparent rhizoboxes enable the performance of repeated measurements of the same roots with a high temporal resolution to quantify static and dynamic characteristics and correlate it to the plant growth performance [18]. Likewise, planar optodes are state-of-the-art technology that can non-invasively measure spatial and temporal dynamics of pH, oxygen (O_2), and carbon (C) dioxide (CO_2) in the soil and in the root-soil interface, which are among the main drivers of processes occurring in the rhizosphere [26–28]. In this scenario, maize has been used as a model plant to investigate rhizosphere processes [29–31].

Within the biogas production scenario, maize has been considered an important energy crop for anaerobic digestion [32]. Conceptually, anaerobic digestion for biogas production consists of the natural breakdown of organic materials by microorganisms in the absence of O_2 to produce biogas as an energy carrier, while the residual by-product is known as digestate [33, 34]. Biogas is a mixture of high-energy methane (CH_4) and CO_2 and is a highly useful source of renewable energy [33–35]. This might be of particular importance in recent times of energy shortages and crises. A wide range of feedstock can be used for biogas production, such as industrial and municipal waste, energy crops, animal manures, and crop residues [36]. Thus, the chemical composition of digestates varies according to the feedstock characteristics and to the digestion conditions as well [33, 34].

Biogas digestate (BD) is a valuable organic fertilizer containing considerable amounts of N, mainly as NH_4^+, among other plant macronutrients such as potassium (K) and phosphorous (P) and micronutrients [37–39]. Therefore, BD has emerged as an alternative to reduce the demand for mineral fertilization, which is associated with a high environmental footprint [40, 41]. In addition, the application of BD in agricultural lands may be considered a sustainable practice of closing the energy and nutrient cycles for environmentally friendly biomass and food production [42]. In the last decades, technological innovation and governmental policies boosted the biogas production sector in Europe [32, 43]. In fact, the number of biogas plants in Europe has tremendously increased from 6227 in 2009 to 17,783 in 2017, with Germany, Italy, and the United Kingdom considered the leaders of biogas production and responsible for consuming 50, 11, and 10%, respectively, of the final gross biogas consumption in the continent [44].

Increasing biogas production inevitably results in increasing BD amounts. Recent estimations suggest that nearly 180 million Mg of BD are produced annually in Europe, where Germany contributes to 48% of this amount [45]. In Germany, most of the biogas produced in 2017 had agricultural residues (44%) and energy crops (48%), mainly maize, as feedstock [44]. Moreover, closely 100% of the so-called agricultural BD, which is composed of a mixture of manure and plants, particularly energy crops, produced in Germany is directly used as fertilizer [45]. Usually, energy crops are co-digested with other materials, such as manure [33]. Although manure alone may result in low biogas yield, it contains great amounts of N, P, and K due to animal excretions, enriching the BD product as fertilizer and concomitantly avoiding the improper disposal of manure in the environment [33]. Yet, as an organic product, BD can contribute to increasing soil organic matter content, water holding capacity, and cation exchange capacity [46], which may enhance soil biological, chemical, and physical attributes [41].

Recently, many studies have been interested in closing the nutrient cycle by adding BD to fields as a renewable form of organic fertilizer [46–50]. Nevertheless, few studies acknowledge that plants are frequently exposed to certain levels of soil compaction in the field, as arable layer compaction, and that mineral and organic fertilizers may undergo distinct dynamics in these circumstances. Furthermore, the interaction between fertilization type and soil compaction may alter root architecture and thus plant performance [11]. Through the investigation of rhizosphere pH dynamics, the uptake of nutrients by plants subjected to such fertilization and soil compaction conditions may be elucidated, allowing insights for the optimization of fertilizers.

We hypothesized that partially replacing regular mineral NPK fertilization with BD fertilization has no drawbacks for the shoot and root parameters of maize seedlings compared to regular mineral NPK fertilization, regardless of the compaction levels tested. In this context, the main purpose of our study was to improve our knowledge regarding sustainable agricultural practices for maize fertilization with BD. Therefore, we investigated through non-invasive techniques (rhizoboxes and planar pH optodes) the performance and root-induced rhizosphere pH variation of juvenile maize in response to ordinary fertilization in Germany with maize silage+chicken manure-derived BD, mineral NPK, and the combination of the two, under arable layer (0–25 cm) compacted substrate and loose substrate. For that, we compared the effects of fertilizer type and substrate compaction condition with respect to: (a) maize shoot development parameters, (b) nutrient content in maize and substrate; (c) maize root architecture; and, (d) maize root-induced rhizosphere pH variation.

2. Materials and methods

2.1 Experimental set-up

A rhizobox experiment was established under greenhouse conditions at IBG-2: Plant Sciences, Forschungszentrum Jülich, in Germany (50.89942° N, 6.39211° E). The experiment was performed in a completely randomized design, considering fertilizer type (Biogas Digestate - BD, mineral NPK fertilizer - NPK, and a mixture of BD and NPK - BD + NPK) and substrate compaction condition (layered compacted substrate and loose substrate) as experimental factors and five replicates, totaling 30 experimental units.

In this study, we aimed to test explicitly the fertilizer effect of BD under the given experimental soil conditions. Therefore, the organic fertilizer (BD) was compared to the mineral NPK fertilizer (control) with reference doses of fertilizer application of 40 and 1.3 Mg ha^{-1}, respectively. The applied doses of BD and NPK fertilizer are in accordance with an ordinary agricultural field application of fresh BD as received from the biogas plant and based on an ordinary agricultural application of 200 kg N ha^{-1}, respectively, which are suitable for maize fertilization [49]. The BD + NPK fertilization consisted of a half BD dose (20 Mg ha^{-1}) + half NPK dose (0.67 Mg ha^{-1}). The BD used in this experiment was a product of the co-digestion of maize silage as the major feedstock and minor amounts of chicken manure, provided by a commercially operating biogas facility (ADRW NaturPower GmbH & Co. Kg Titz-Ameln, Germany). The BD contained: 7.2% dry matter, 0.53% total N, 0.32% NH_4^+, 0.14% P, 0.68% K, 0.037% Mg, 0.16% Ca, 0.03% S, 5.3% organic matter, C:N ratio 6, and pH in $CaCl_2$ 8.2 (all values referring to fresh weight). The mineral NPK fertilizer (Compo Rasendünger, Compo GmbH, Münster, Germany) contained: 15% N (1% NO_3^-, 9.5% ammonia, 4.5% isobutylidenediurea), 5% P, 8% K, and 3% Mg. The amounts of N, P, and K present in the applied doses of BD, NPK, and BD + NPK fertilizer were calculated in kg ha^{-1} and are described in **Table 1**. Application of NPK and BD resulted in similar N and P doses (**Table 1**). The higher K dose derived from BD compared to NPK was considered for the experiment since this type of BD (maize

Fertilizer	N (kg ha^{-1})	P (kg ha^{-1})	K (kg ha^{-1})	N:P:K
BD[1]	212	56	272	1.00:0.26:1.28
BD + NPK	206	61	189	1.00:0.30:0.92
NPK[2]	200	67	107	1.00:0.33:0.53

[1]N, P and K content in the full BD fertilizer applied dose was calculated based on 0.53% total N, 0.14% P, and 0.68% K in accordance with an agricultural field application of 40 t ha^{-1} of digestate.
[2]N, P and K content in the full NPK fertilizer applied dose was calculated based on 15% total N, 5% P, and 8% K in accordance with an agricultural field application of 200 kg N ha^{-1}.

Table 1.
Nitrogen (N), phosphorus (P) and potassium (K) content and the N:P:K ratio in the applied doses of Biogas Digestate (BD), Biogas Digestate+mineral NPK fertilizer (BD + NPK) and mineral NPK fertilizer (NPK).

silage and chicken manure feedstocks) is ordinarily applied in place of mineral NPK fertilizer in Germany [44, 45].

Maize plants were grown in a mixture of 50% (v/v) arable field soil (Endogleyic Stagnosol, with a silty-loamy texture containing 1.1% C; 0.1% N; 0.1% P; 1.6% K; pH in $CaCl_2$: 6.3) collected from 0 to 30 cm depth near Jülich municipality, and 50% (v/v) black peat potting substrate (Einheitserde Null; containing 19.3% C; <0.01% N; 0.7% P; <0.01% K; pH in $CaCl_2$: 6.0). Therefore, the growth medium was called substrate. The substrate was manually sieved to a maximal aggregate size of 0.5 cm. The utilization of dark peat improves the visual contrast between plant roots and the growth medium in the rhizoboxes, enhancing the detection of roots during the non-invasive root growth analysis.

The fertilizers were homogeneously mixed into the substrate using an end-over-end mixer for 15 min. The rhizoboxes (dimensions: 60 cm height x 30 cm width x 2 cm depth) were filled with the substrate-fertilizer mixture, and thereafter the compaction protocol (see layered compacted and loose substrate rhizoboxes section) was followed. Subsequently, two maize seeds were sown in the top center of the rhizoboxes at a depth of 2 cm in relation to the substrate surface. One day after seedling emergence, the smallest seedling was manually removed from the rhizobox. Plants were watered with 100 mL of deionized water three times per week to maintain a minimum of 30% volumetric water content in the substrate during the experiment. During cultivation, the rhizoboxes were kept at an inclination angle of 45° to force the roots to grow along the transparent plate of the rhizobox [51]. The transparent plates were facing downwards on the horizontal plane and were covered with lightproof material (black plate) throughout the experiment to avoid light disturbances to root growth and to planar pH optodes signal (see pH measurements in Section 2.3). Plants were grown for 18 days after germination (DAG) under controlled conditions (16 h light per day, day/night temperature of 22/17°C, and 60% relative humidity).

2.2 Layered, compacted, and loose substrate rhizoboxes

A protocol described by Nagel et al. [18] was adapted to obtain rhizoboxes with different levels of compaction, named layered compacted substrate (C/L) and loose substrate (L/L), as illustrated in **Figure 1**. The C/L rhizoboxes contained a top layer (0–25 cm depth) of compacted substrate followed by a bottom layer (25–55 cm depth) of loose substrate. The bottom layer of the loose substrate was obtained by pouring gradually 1 L portions of the substrate into the rhizobox. Subsequently, the compacted top layer of 25 cm depth was obtained by pouring gradually 1 L portions of the substrate into the rhizobox, and then each 1 L portion was compressed by using a custom-built compaction frame. The compression was applied with a manual pallet forklift by lifting individual rhizoboxes against the frame while pressure was applied to the substrate surface using a wooden plank. Each compressed 1 L portion of substrate received an equivalent pressure of 1.2 kg cm^{-2}. The L/L rhizoboxes contained a continuous loose layer of 55 cm depth that was obtained by pouring gradually 1 L portions of substrate into the rhizobox. Two drainage holes (diameter of 0.8 cm) at the bottom of the rhizoboxes allowed sufficient substrate drainage.

Three additional replicates of C/L and L/L rhizoboxes were destined to determine the substrate dry bulk density and penetration resistance. To determine the substrate bulk density, the frontal rhizobox plates (transparent) were opened (with minimal disturbance) to collect substrate samples at 10 and 50 cm depth using sampling

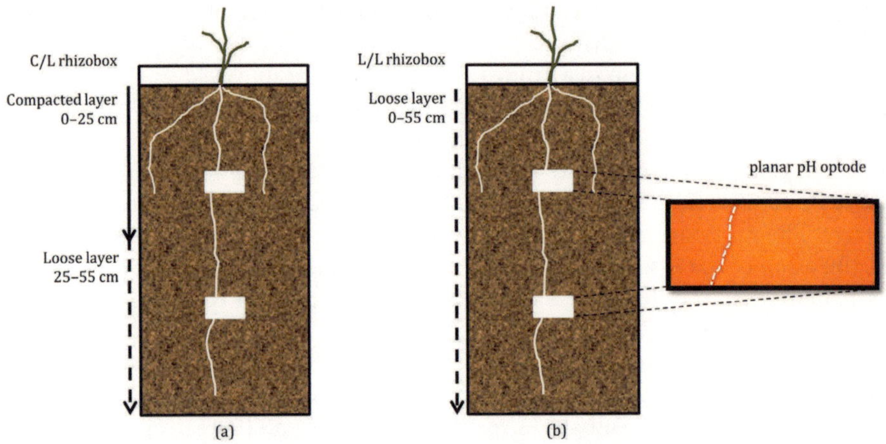

Figure 1.
Drawings illustrate: (a) layered compacted (C/L) rhizobox: top compacted layer (0–25 cm depth; 1.0 MPa, 0.84 g cm^{-3} at 10 cm depth) followed by loose layer (25–55 cm depth; 0.66 g cm^{-3} at 50 cm depth), and (b) loose (L/L) rhizobox: continuous loose layer (0–55 cm depth; 0.04 MPa, 0.62 and 0.61 g cm^{-3} at 10 and 50 cm depth). White rectangles represent planar pH optodes positioned at 15 and 35 cm depth.

cylinders (volume: 10.6 mL). The samples were dried for 48 h at 105°C and weighed. The substrate dry bulk density was obtained by dividing the dry substrate mass by the volume of the sampling cylinder. The substrate penetration resistance was measured with a hand penetrometer for the top layers (Eijkelkamp Agrisearch Equipment, Giesbeek, Netherlands). The penetrometer had a 30° cone angle, an 8 mm maximum cone diameter, and a penetration depth of 10 cm and was fitted with 50-N steel springs. The C/L substrate presented densities of 0.84 ± 0.04 and 0.66 ± 0.05 g cm^{-3} at 10 cm and 50 cm depths respectively, and penetration resistance of 1.0 MPa at 10 cm depth. The L/L substrate presented bulk densities of 0.62 ± 0.04 and 0.61 ± 0.04 g cm^{-3} at 10 cm and 50 cm depths respectively, and penetration resistance of 0.04 MPa at 10 cm depth.

2.3 pH measurements via planar optode technique

The planar pH optode is a sensor foil with embedded fluorescent molecules that emit a characteristic pattern of fluorescence after excitation with light, which allows measurements of pH *in vivo* and *in situ* [52]. The pattern of fluorescence depends on the concentration of the analyte. In the present study, protons were the analyte, and planar pH optodes were used to monitor non-invasively and quantitatively the spatial and temporal dynamics of rhizosphere pH over the experimental period of 18 days. The relatively short experimental period can provide relevant information on maize rhizosphere pH changes and associated nutrition as younger maize plants have considerably higher root exudation activity than older ones [53], being assumingly more prone to detection by the used optodes. As previously described by Blossfeld et al. [52], optode measurements use the ratiometric approach, which is based on the change of the emission spectrum of the planar optodes depending on the analyte concentration. In our study, the ratio of red to green in the emitted fluorescence response was recorded in the respective color channels of an RGB (red, green, blue) chip. The RGB chip of the camera can capture the fluorescence signals in one single image, and the subsequent data analysis by the software creates the ratio of the red

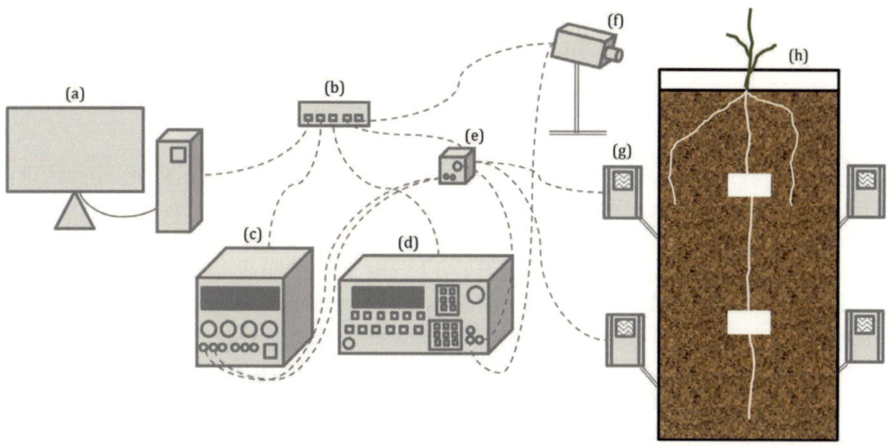

Figure 2.
Schematic drawing of the setup of VisiSens TD1 system, including: (a) computer system, (b) switch, (c) power supply, (d) function generator, (e) strobe controller, (f) color camera including objective, (g) 4 light sources, (h) rhizobox with upper (15 cm depth) and lower (35 cm depth) planar pH optodes. Dotted lines represent the electric, trigger, and network cables.

and the green channel (i.e., R-value), creating a two-dimensional map of the quantity of the measured parameter.

The VisiSens TD1 system (PreSens GmbH, Regensburg, Germany) was used to process the pH data. The system relies on a camera device, a power supply, a function generator, a strobe controller, a light source for pH measurements, and a software to identify pH changes in the root zone through the pH-sensitive planar optodes (product code SF-HP5-OIW; PreSens GmbH) (**Figure 2**). In this study, since maize sowing, one image per day of the optode sensors was taken using a distance of 12.5 cm between the transparent plate of the rhizobox where the planar pH optode was positioned and the camera, resulting in a pixel resolution of 1292 x 964. Every pixel carries discrete quantitative analyte information. The software ImageJ was then used to compute the quantitative maps from the raw sensor response images. Detailed information on high-resolution color radiometric planar optode imaging approaches was previously described by Larsen et al. [54].

Calibration of the pH planar optodes was performed prior to and at the end of the experimental phase in accordance with Blossfeld and Gansert [26]. For that, two stock solutions of phosphate buffers with the same ionic strength (solution A: 0.1 mol L^{-1} of K_2HPO_4 and solution B: 0.1 mol L^{-1} of KH_2PO_4 + 0.2 mol L^{-1} of NaCl) were used to produce seven calibration buffer solutions with different pH values (4.6, 5.0, 5.5, 6.0, 6.5, 7.0, and 7.5) by mixing different volumes of the stock solutions. A replicate of the planar pH optode used in the rhizoboxes was installed in the calibration chamber. The calibration chamber was then subsequently filled with each of the pH buffer solutions, and the change of the R-values of the planar pH optode was monitored from outside the system every minute over a period of 10 min, at which time the R-value was constant. The resulting R-values were used as input data for a Boltzmann fitting curve. The resulting fitting curve was then used to calculate the pH value.

For all images obtained with the same sensor throughout the experimental period, a fixed bulk substrate area (without roots) and a fixed root area (rhizosphere) were selected and analyzed using the software ImageJ to create temporal and quantitative maps of pH. Rhizosphere pH variation promoted by maize roots was based on rhizosphere pH and bulk substrate pH (Δ rhizosphere pH) obtained in each planar pH optode sensor and was calculated as described in Eq. (1):

$$\Delta\,\text{rhizosphere}\ pH = bulk\ substrate\ pH - rhizosphere\ pH \tag{1}$$

Rhizosphere pH variation over the experimental period, shown as alkalization and acidification pH units, are presented as positive and negative values, respectively.

Prior to filling the rhizoboxes with substrate, the planar pH optode (dimensions: 3 cm height x 8 cm width) was fixed to the inner surface of the front (transparent) plate of the rhizobox at 15 and 35 cm depth (in relation to substrate surface level) using a thin layer of silicon grease as an adhesive (**Figure 1**). The pre-installation of a planar optode in a fixed area of the rhizobox may increase the chances of roots growing outside the installation area. However, pre-installed planar optodes are expected to reduce overall disturbances to the system as compared to opening the rhizobox after roots have grown attached to the front plate of the rhizobox.

2.4 Root system development measurements

Images of the root systems were firstly acquired at 4 DAG and afterward, every 2 to 4 days during the experimental period using a single-lens reflex camera (SLR, 10 Mpx, with a 28 mm electro-focus lens EFS, EOS digital 400D; Canon USA, Inc., Melville, NY, USA). Images were acquired within a custom-made photo box to avoid light interferences in the images. The resolution of the acquired images was adequate for the detection of main and lateral roots. The image analysis software GROWSCREEN-Root [18] was used to manually draw the visible roots growing attached to the transparent plate of the rhizobox and to quantify the visible main roots length, total lateral root length, root system width, and rooting depth.

2.5 Maize development and substrate parameters

At 18 DAG, the photosynthetic assimilation was measured on the third youngest leaf of all individual plants using a soil plant analysis development chlorophyll meter (SPAD-502, Konika-Minolta, Marunouchi, Japan). At harvest, maize plants had four leaves, and the investigated leaf was fully expanded. Subsequently, the shoot (stem and leaves) was harvested at substrate level, and leaf area measurements were determined using a leaf area meter (LI-3100 Area meter, LI-COR Inc., Lincoln, NE, USA) to calculate the specific leaf area (SLA), as described in Eq. (2):

$$SLA = leaf\ area\left(m^2\right) \div leaf\ dry\,weight\left(kg^{-1}\right) \tag{2}$$

After harvesting, the rhizoboxes were opened and substrate samples (100 g) were taken from all replicates at 20 and 45 cm depth. Thereafter, roots were separated from the substrate by carefully flushing with water. Shoot and root biomass was dried at 70°C and substrate samples were dried at 30°C until constant weight was reached. Maize root:shoot ratio was calculated based on the biomass dry weight. Shoot, root, and soil samples were milled using a swing mill (Retsch MM200, Retsch GmbH,

Haan, Germany), homogenized and analyzed for C and N content using an elemental analyzer (vario Max CNS, Elementar GmbH, Langenselbold, Germany) and for P and K content using the inductively coupled plasma-optical emission spectrometer (ICP-OES, iCAP7600, Thermo Scientific, Dreieich, Germany).

2.6 Statistical analysis

Normality of data was tested using the Shapiro-Wilk test, which revealed a normal distribution of the data. In order to test treatment effects on response variables, the data were submitted to analysis of variance (ANOVA). A post-hoc Tukey test was applied to compare means of treatments when ANOVA indicated significant effects of experimental factors (fertilizer type, substrate compaction condition, and the interaction between the two). Statistical significance was set as $p < 0.05$.

3. Results

The p values of ANOVA testing the effect of fertilizer type, substrate compaction condition, and interaction between the two on the response variables at harvest (18 DAG) are shown in **Table 2**.

3.1 Maize shoot and root dry weight

Maize germination occurred similarly across treatments 3 days after sowing. Means of maize shoot and root dry weight at harvest (18 DAG) are shown in **Figure 3**. Maize plants presented shoot dry weight ranging from 416 (NPK L/L) to 550 mg plant^{-1} (NPK C/L). ANOVA revealed neither significant interaction between fertilizer type and compaction condition nor significant effect of the main factors for shoot dry weight (**Table 2**). These results indicate that BD fertilization promoted similar maize shoot biomass compared to NPK and BD + NPK fertilization, regardless of compaction condition.

Means of root dry weight obtained from 0 to 25 cm and 25 to 55 cm depth ranged from 78 (BD C/L) to 107 (BD + NPK C/L) mg plant^{-1} and from 14 (BD + NPK L/L) to 31 (NPK C/L) mg plant^{-1}, respectively (**Figure 3**), and were affected neither by the interaction between fertilizer type and compaction condition nor by the main factors (**Table 2**). Nevertheless, when considering total root dry weight (0–55 cm depth), significant interaction between fertilizer type and compaction condition occurred (**Table 2**). Under C/L, total root dry weight varied from 96 (BD) to 133 mg plant^{-1} (NPK) (**Figure 3**). In this case, the total root dry weight observed in NPK and NPK + BD treatment did not differ from each other, and it was about 38 and 39% higher than that in BD, respectively (**Figure 3**). Under this compaction condition, the proportion of the total root dry weight (0–55 cm depth) found at 0–25 and 25–55 cm depths slightly ranged from 77 to 81% and 19 to 23%, respectively, across fertilizer types (**Figure 3**). Under L/L, total root dry weight did not differ between fertilizer types and ranged from 105 (NPK) to 109 mg plant^{-1} (BD) (**Figure 3**), indicating that root dry weight promoted by BD was equivalent to that promoted by NPK and NPK + BD. Under this compaction condition, the proportion of the total root dry weight (0–55 cm depth) found at 0–25 and 25–55 cm depth varied from 80 to 86% and from 13 to 20%, respectively (**Figure 3**).

Variables	**F**	**C**	**F*C**
Shoot parameters			
Dry weight	> 0.05	> 0.05	> 0.05
Number of leaves	> 0.05	> 0.05	> 0.05
Specific leaf area	> 0.05	> 0.05	> 0.05
SPAD readings	> 0.05	> 0.05	> 0.05
root:shoot ratio	**0.037**	> 0.05	> 0.05
C content	> 0.05	> 0.05	> 0.05
N content	> 0.05	> 0.05	> 0.05
P content	**<0.001**	**<0.001**	> 0.05
K content	> 0.05	> 0.05	> 0.05
Root parameters			
Dry weight (0–25 cm)	> 0.05	> 0.05	> 0.05
Dry weight (25–55 cm)	> 0.05	> 0.05	> 0.05
Total dry weight (0–55 cm)	> 0.05	> 0.05	**0.017**
C content	> 0.05	**0.036**	> 0.05
N content	> 0.05	> 0.05	**0.013**
P content	**<0.001**	**<0.001**	> 0.05
K content	**0.029**	**0.007**	**<0.001**
Visible main roots length	> 0.05	> 0.05	**0.001**
Visible total lateral root length	**<0.001**	> 0.05	> 0.05
Visible root system width	> 0.05	> 0.05	> 0.05
Visible rooting depth	> 0.05	> 0.05	**0.046**
Soil parameters			
C content (0–25 cm)	> 0.05	> 0.05	> 0.05
N content (0–25 cm)	**<0.001**	> 0.05	> 0.05
P content (0–25 cm)	> 0.05	> 0.05	> 0.05
K content (0–25 cm)	**0.022**	**0.013**	> 0.05
C content (25–55 cm)	**0.033**	> 0.05	> 0.05
N content (25–55 cm)	**0.002**	> 0.05	> 0.05
P content (25–55 cm)	> 0.05	> 0.05	> 0.05
K content (25–55 cm)	**<0.001**	> 0.05	> 0.05

Table 2.
*ANOVA (2 x 3 factorial) results indicating effects of fertilizer type (F), substrate compaction condition (C) and their interaction (F*C) on shoot, root and soil response variables at harvest.* $p < 0.05$ *is written in bold.*

3.2 Maize development parameters

Means of maize development parameters such as number of leaves, SLA, SPAD readings, and root:shoot ratio at harvest (18 DAG) are presented in **Table 3**. Maize plants at harvest presented four fully expanded leaves in all treatments. The SLA

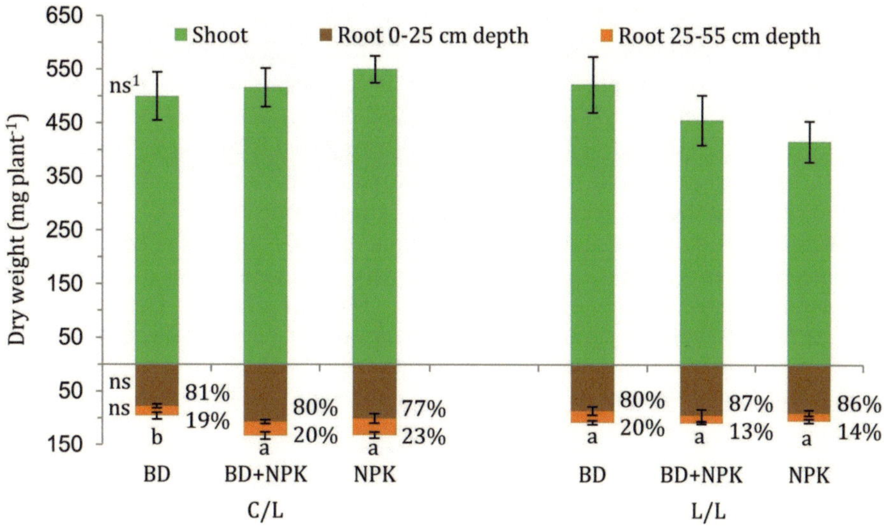

Figure 3.
Dry weight of maize shoot and root at harvest as affected by fertilizer type (Biogas Digestate - BD, Biogas Digestate+mineral NPK fertilizer - BD + NPK, mineral NPK fertilizer - NPK) and compaction condition (layered compacted substrate - C/L, loose substrate - L/L). Data are mean (n = 5) and error bars are standard error. Means of total root dry weight (0–55 cm depth) with different letter within compaction condition (C/L, L/L) are statistically different according to Tukey's test at p < 0.05. [1]ns = mean of shoot weight and root weight at 0–25 and 25–55 cm depth did not differ statistically according to Tukey's test at p < 0.05. Percentage numbers are the proportion of the total root dry weight (0–55 cm depth) at 0–25 and 25–55 cm depth.

Fertilizer	C/L	L/L	Mean[1]	C/L	L/L	Mean[1]
	Number of leaves			SPAD readings		
BD	4(0.0)	4(0.0)	4	44.5(1.3)	45.1(1.7)	44.8
BD + NPK	4(0.0)	4(0.0)	4	44.1(0.9)	40.6(0.8)	42.4
NPK	4(0.2)	4(0.0)	4	43.2(1.2)	40.9(0.7)	42.1
Mean[2]	4	4		44.0	42.2	
	Specific leaf area (m^2 kg^{-1})			Root:shoot ratio		
BD	37.5(2.0)	39.5(3.8)	38.5	0.198(0.038)	0.211(0.030)	0.204 B
BD + NPK	37.5(2.4)	39.7(1.4)	38.6	0.261(0.035)	0.239(0.030)	0.250 AB
NPK	35.5(2.0)	37.7(1.2)	36.6	0.243(0.034)	0.264(0.079)	0.253 A
Mean[2]	36.9	39.9		0.234	0.238	

[1]*Mean of fertilizer type (n = 10).*
[2]*Mean of compaction condition (n = 15).*

Table 3.
Maize number of leaves, specific leaf area, SPAD readings and root:shoot ratio at harvest as affected by fertilizer type (Biogas Digestate - BD, Biogas Digestate+mineral NPK fertilizer - BD + NPK, mineral NPK fertilizer - NPK) and substrate compaction condition (layered compacted substrate - C/L, loose substrate - L/L. Data are mean followed by standard error within parentheses. Means with the same lowercase letter for fertilizer type within compaction condition (n = 5) or same uppercase letter for compaction condition mean (n = 15) and fertilizer type mean (n = 10) are not statistically different according to Tukey's test at p < 0.05.

means ranged from 35.5 (NPK C/L) to 39.7 $m^2 kg^{-1}$ (BD + NPK L/L), and SPAD reading means varied from 40.6 (BD + NPK L/L) to 45.1 (BD L/L) (**Table 3**). ANOVA revealed neither a significant effect of the interaction between fertilizer type and compaction conditions nor significant effects of the main factors for number of leaves, SLA, and SPAD readings (**Table 2**), indicating that BD fertilization enabled similar maize development up to 18 days compared to NPK and BD + NPK with respect to such parameters, regardless of compaction condition.

Regarding root:shoot ratio, ANOVA showed neither significant effects of the interaction between fertilizer type and compaction condition nor significant effects of compaction condition main factor, whereas significant effects of fertilizer type occurred (**Table 2**). The root:shoot ratio fertilizer mean (n = 10) was significantly higher in NPK (0.253) than in BD (0.204), while the root:shoot ratio in BD + NPK (0.250) did not differ from the other two fertilizers (**Table 3**).

3.3 Shoot and root chemical analyses

Means of C, N, P, and K content in maize shoot and root at harvest (18 DAG) are presented in **Table 4**. Maize shoot C content ranged from 389.5 (NPK L/L) to 396.0 g kg^{-1} (NPK C/L), N content from 45.9 (BD L/L) to 49.2 g kg^{-1} (BD + NPK

	Shoot			Root		
Fertilizer	**C/L**	**L/L**	**Mean[1]**	**C/L**	**L/L**	**Mean[1]**
BD	395.1(1.6)	394.8(2.5)	395.0	386.6(6.2)	393.2(1.5)	389.9
BD + NPK	390.5(0.4)	391.7(1.2)	391.1	380.2(0.9)	385.0(3.3)	382.6
NPK	396.0(3.7)	389.5(1.3)	395.0	381.2(8.9)	396.0(2.1)	388.6
Mean[2]	393.9	392.0		382.7 B	391.4 A	
BD	49.0(0.4)	45.9(2.6)	47.5	29.6(0.6) a	29.0(0.1) b	29.3
BD + NPK	49.2(0.5)	47.9(0.6)	48.6	29.1(0.4) a	32.5(1.2) a	30.8
NPK	46.6(2.1)	47.7(0.8)	47.2	30.0(1.3) a	28.6(0.2) b	29.3
Mean[2]	48.3	47.2		29.6	30.0	
BD	5.2(0.2)	6.3(0.2)	5.8 B	3.1(0.1)	3.7(0.1)	3.4 B
BD + NPK	6.7(0.4)	8.2(0.5)	7.5 A	3.1(0.2)	4.0(0.1)	3.6 B
NPK	7.9(0.7)	9.2(0.4)	8.6 A	4.2(0.3)	4.4(0.1)	4.3 A
Mean[2]	6.6 B	7.9 A		3.6 B	4.1 A	
BD	65.2(1.2)	65.0(1.2)	65.1	16.4(0.5) b	31.0(4.0) a	23.7
BD + NPK	65.7(0.7)	66.2(1.6)	66.0	16.1(0.5) b	22.0(1.4) ab	19.1
NPK	64.7(1.6)	63.6(0.8)	64.2	27.2(1.8) a	20.9(0.7) b	24.0
Mean[2]	65.2	65.0		19.9	24.6	

[1]Mean of fertilizer type (n = 10).
[2]Mean of compaction condition (n = 15).

Table 4.
Carbon (C), nitrogen (N), phosphorous (P) and potasium (K) content in maize shoot and root at harvest as affected by fertilizer type (Biogas Digestate - BD, Biogas Digestate+mineral NPK fertilizer - BD + NPK, mineral NPK fertilizer - NPK) and compaction condition (layered compacted substrate - C/L, loose substrate - L/L). Data are mean followed by standard error within parentheses. Means with the same lowercase letter for fertilizer type within compaction condition (n = 5) or same uppercase letter for compaction condition mean (n = 15) and fertilizer type mean (n = 10) are not statistically different according to Tukey's test at $p < 0.05$.

C/L), P content from 5.2 (BD C/L) to 9.2 g kg^{-1} (NPK L/L), and K content from 63.6 (NPK L/L) to 66.2 g kg^{-1} (BD + NPK L/L) across the different treatments (**Table 4**). ANOVA revealed neither significant interaction between fertilizer type and compaction condition nor significant effect of the main factors for shoot C, N, and K contents (**Table 2**).

Maize shoot P content was significantly affected by both factors fertilizer type and compaction condition, but not by the interaction of both (**Table 2**). Regardless of compaction condition, shoot P content fertilizer mean (n = 10) promoted by BD + NPK (7.5 g kg^{-1}) and NPK (8.6 g kg^{-1}) did not differ statistically from each other, but it was 29 and 33% higher than that of BD (5.8 g kg^{-1}) (**Table 4**). These results point out that P nutrition of juvenile maize shoot promoted by BD alone was reduced in relation to the fertilization sources containing mineral fertilizer (NPK). Independently of fertilizer type, shoot P content compaction condition mean (n = 15) in L/L (7.9 g kg^{-1}) was 20% higher than in C/L (6.6 g kg^{-1}) (**Table 4**).

Maize root C content ranged from 380.2 (BD + NPK C/L) to 396.0 g kg^{-1} (NPK L/L), N content from 28.6 (NPK L/L) to 32.5 g kg^{-1} (BD + NPK L/L), P content from 3.1 (BD and BD + NPK C/L) to 4.4 g kg^{-1} (NPK L/L), and K content from 16.1 (BD + NPK C/L) to 31.0 g kg^{-1} (BD L/L) across the different treatments. In general, treatments apparently affected more remarkably the elemental content in maize roots than in shoots. For instance, ANOVA revealed that N and K content in maize roots were affected by the interaction between fertilizer type and compaction condition, that C and P content in maize roots were affected by compaction condition, and that P content in maize roots was also affected by fertilizer type (**Table 2**).

With respect to root C content, mean of compaction condition (n = 15) was slightly (2.3%) higher in L/L (391.4 g kg^{-1}) than in C/L (382.7 g kg^{-1}) (**Table 4**). Regarding root N content, the fertilizer type did not differ from each other within C/L condition, while within L/L condition, root N content promoted by BD + NPK (32.5 g kg^{-1}) was 12 and 14% higher than that promoted by BD (29.0 g kg^{-1}) and NPK (28.6 g kg^{-1}), respectively, which did not differ from each other (**Table 4**). Overall, these results demonstrate that BD promoted similar root N content in relation to NPK, regardless of compaction condition, and yet that under L/L conditions, the mixture of BD with NPK (BD + NPK) incremented the root N content in relation to regular NPK fertilization.

Independently on compaction condition, root P content fertilizer mean (n = 10) resulting from NPK application (4.3 g kg^{-1}) was 27 and 19% higher compared to that conferred by BD (3.4 g kg^{-1}) and BD + NPK (3.6 g kg^{-1}), respectively, which did not differ from each other (**Table 4**). Still, regarding root P content, the mean of compaction condition (n = 15) was 12% higher in L/L (4.1 g kg^{-1}) than in C/L (3.6 g kg^{-1}) (**Table 4**). The pattern of root P content in response to the sources of variation, particularly to fertilizer type, is conformable to that observed for shoot P content. Together, such patterns suggest that maize fertilization with the BD type tested in this study should be complemented by other P sources, preferably highly soluble P fertilizers, in order to avoid possible juvenile maize P deficiency in relation to regular mineral fertilization.

Root K content was responsive to the interaction between fertilizer type and compaction condition. Within C/L condition, root K content in response to NPK fertilization (27.2 g kg^{-1}) was 66 and 70% higher compared to that promoted by BD (16.4 g kg^{-1}) and BD + NPK (16.1 g kg^{-1}), respectively, which did not differ from each other (**Table 4**). Contrastingly, within L/L condition, root K content promoted by BD application (31.0 g kg^{-1}) was 48% higher than that resulting from NPK (20.9 g kg^{-1})

application, and root K content due to BD + NPK amendment (22.0 g kg^{-1}) did not differ from the other fertilizers (**Table 4**).

3.4 Substrate chemical analyses at harvest

Means of C, N, P, and K content in substrate samples collected at 0–25 and 25–55 cm depths at maize harvest are shown in **Table 5**. Substrate C content at 0–25 cm depth ranged from 60.5 (NPK L/L) to 65.0 g kg^{-1} (BD + NPK L/L), and at 25–55 cm from 56.6 (NPK L/L) to 65.4 g kg^{-1} (BD L/L) (**Table 5**). In general, substrate C content tended to be lower in NPK treatments compared to that with BD. However, ANOVA revealed that such a tendency was statistically significant only at 25–55 cm depth, where substrate C content was significantly affected by fertilizer type (**Table 2**). At this depth, C content fertilizer type mean (n = 10) in BD (63.9 g kg^{-1}) was 9% higher compared to NPK (58.9 g kg^{-1}), whereas BD + NPK mean (62.4 g kg^{-1}) did not differ from the other two fertilization types (**Table 5**).

Substrate N content at 0–25 cm depth varied from 2.1 (NPK L/L) to 2.6 g kg^{-1} (BD + NPK C/L and L/L), and at 25–55 cm depth it ranged from 2.3 (NPK L/L) to

	0–25 cm depth			25–55 cm depth		
Fertilizer	C/L	L/L	Mean[1]	C/L	L/L	Mean[1]
BD	64.4(1.9)	62.2(2.9)	63.3	62.4(1.2)	65.4(3.0)	63.9 A
BD + NPK	64.7(2.4)	65.0(1.6)	64.8	61.2(1.0)	63.6(2.1)	62.4 AB
NPK	64.0(1.8)	60.5(1.0)	62.3	61.1(1.6)	56.6(1.2)	58.9 B
Mean[2]	64.4	62.6		61.6	61.9	
BD	2.4(0.1)	2.4(0.1)	2.4 AB	2.4(0.0)	2.4(0.1)	2.4 A
BD + NPK	2.6(0.1)	2.6(0.1)	2.6 A	2.4(0.0)	2.5(0.0)	2.5 A
NPK	2.4(0.0)	2.1(0.1)	2.3 B	2.3(0.1)	2.3(0.0)	2.3 B
Mean[2]	2.5	2.4		2.4	2.4	
BD	815.9(26.1)	790.7(10.5)	803.3	840.9(12.9)	778.2(7.9)	809.6
BD + NPK	825.5(14.5)	813.7(20.1)	819.6	817.6(11.9)	835.8(20.6)	826.7
NPK	841.8(26.2)	835.1(22.2)	838.4	819.6(21.5)	820.2(13.7)	819.9
Mean[2]	827.7	813.2		826.0	811.4	
BD	13.0(0.3)	15.0(0.7)	14.0 AB	14.0(0.2)	14.2(0.1)	14.1 B
BD + NPK	14.8(0.4)	15.0(0.2)	14.9 A	14.8(0.1)	15.4(0.2)	15.1 A
NPK	13.1(0.2)	14.0(0.7)	13.6 B	14.9(0.2)	14.8(0.3)	14.6 A
Mean[2]	13.7 B	14.7 A		14.6	14.8	

[1]*Mean of fertilizer type (n = 10).*
[2]*Mean of compaction condition (n = 15).*

Table 5.
Carbon (C), nitrogen (N), phosphorous (P) and potassium (K) content in the substrate at 0–25 and 25–55 cm depth at harvest as affected by fertilizer type (Biogas Digestate - BD, Biogas Digestate+mineral NPK fertilizer - BD + NPK, mineral NPK fertilizer - NPK) and compaction condition (layered compacted substrate - C/L, loose substrate - L/L). Data are mean followed by standard error within parentheses. Means with the same lowercase letter for fertilizer type within compaction condition (n = 5) or same uppercase letter for compaction condition mean (n = 15) and fertilizer type mean (n = 10) are not statistically different according to Tukey's test at $p < 0.05$.

2.5 g kg^{-1} (BD + NPK L/L) (**Table 5**). For both depths, ANOVA revealed that substrate N content was significantly affected only by fertilizer type (**Table 2**). At 0–25 cm depth, substrate N content fertilizer mean (n = 10) in BD + NPK (2.6 g kg^{-1}) was 13% higher than that in NPK (2.3 g kg^{-1}), and BD mean (2.4 g kg^{-1}) was indistinct from the others (**Table 5**). Similarly, at 25–55 cm depth, substrate N content fertilizer mean (n = 10) decreased as BD + NPK = BD > NPK (**Table 5**).

Means of substrate P content at 0–25 cm depth varied from 790.7 (BD L/L) to 841.8 mg kg^{-1} (NPK C/L), while at 25–55 cm depth the range was 778.2 (BD L/L) and 840.9 mg kg^{-1} (BD C/L) (**Table 5**). For both depths, ANOVA revealed that substrate P content was affected neither by the interaction between compaction condition and fertilizer type nor by these two main factors (**Table 2**).

Means of substrate K content at 0–25 cm depth ranged from 13.0 (BD C/L) to 15.0 g kg^{-1} (BD and BD + NPK L/L), and at 25–55 cm depth it varied from 14.0 (BD C/L) to 15.4 g kg^{-1} (BD + NPK L/L) (**Table 5**). For both depths, ANOVA revealed that substrate K content was significantly affected by fertilizer type (**Table 2**). Yet, at 0–25 cm depth, compaction condition also significantly affected substrate K content (**Table 2**). Contrasting patterns of fertilizer mean (n = 10) were observed among substrate depths. At 0–25 cm depth, substrate K content fertilizer mean (n = 10) in BD + NPK was 10% higher than that in NPK, and BD (14.0 g kg^{-1}) did not differ from the other two treatments (**Table 5**). Still at this depth, K content compaction condition mean (n = 15) in L/L was 7% higher in comparison to C/L (**Table 5**). At 25–55 cm depth, substrate K content fertilizer mean (n = 10) in BD + NPK and NPK was slightly higher (approx. 5%) than in BD (14.1 g kg^{-1}), respectively (**Table 5**).

3.5 Root system development

Means of visible main roots length, total lateral root length, root system width, and rooting depth over the experimental period, as well as statistical significance between treatments for these parameters at harvest (18 DAG), are shown in **Figure 4**. ANOVA revealed that visible main root length and rooting depth were significantly affected by the interaction between fertilizer type and compaction condition, while visible total lateral root length was significantly affected only by fertilizer type, and visible root system width was affected neither by the interaction between compaction condition and fertilizer type nor by the two main factors (**Table 2**).

Means of visible main roots length under C/L condition for BD, BD + NPK and NPK were 108.5, 104.3, and 84.2 cm, respectively, and did not differ from each other (**Figure 4a**). Differently, within L/L condition, the visible main roots length mean in NPK (107.6 cm) was significantly higher (49%) than in BD (72.2 cm), whereas BD + NPK mean (93.1 cm) did not differ from the others (**Figure 4b**). Regarding visible total lateral root length, the mean values (n = 10) of NPK (198.1 cm) and BD (187.0 cm) did not differ from one another and were 60% and 51% higher in comparison with BD + NPK (124.2 cm), respectively (see **Figure 4d** and its statistical footnote). As a general trend for visible main root length and total lateral root length of contrasting fertilizer treatments (BD versus NPK), it was verified that BD application resulted in a significant reduction of main root length (in L/L) (**Figure 4b**) and a similar total lateral root length in relation to NPK (**Figure 4d**).

Treatments did not affect the width of maize root system either under C/L (**Figure 5a**) or under L/L (**Figure 5b**). Maize visible rooting depth under C/L condition reached depths of 50.1, 51.3, and 54.2 cm in NPK, BD + NPK and BD treatments, respectively, and means were not statistically different (**Figure 5c**). Whereas, within

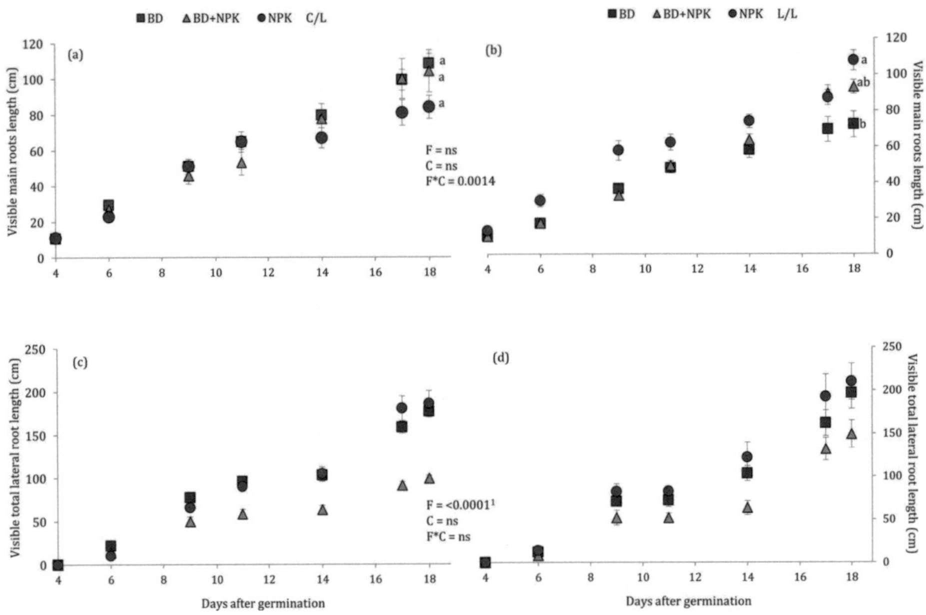

Figure 4.
*Effect of fertilizer type (Biogas Digestate - BD, Biogas Digestate+mineral NPK fertilizer - BD + NPK, mineral NPK fertilizer - NPK) and compaction condition (layered compacted substrate - C/L, loose substrate - L/L) on maize visible main roots length (a, b) and total lateral root length (c, d) over days after germination. Data are mean ($n = 5$) and error bars are standard error. Means with the same letter within compaction condition (C/L, L/L) at 18 days after germination are not statistically different according to Tukey's test at $p < 0.05$. F = effect of fertilizer; C = effect of compaction; F*C = interaction between fertilizer and compaction; ns = not significant.*
[1]The effect of fertilizer type ($n = 10$) was statistically significant ($p < 0.0001$) on visible total lateral root length (cm) as it follows: NPK (198.1 a), BD (187.0 a), BD + NPK (124.2 b).

L/L condition, roots grew 23% deeper in NPK (54 cm) than in BD + NPK (44 cm), and that in BD (48.1 cm) did not differ from the other two treatments (**Figure 5d**). Together these results indicate that juvenile maize root system exploration in depth and width promoted by BD was equivalent to that promoted by NPK, irrespective of compaction condition.

3.6 Rhizosphere and bulk substrate pH over days after germination

Three of the five replicates of each treatment contained planar pH optodes sensors positioned at 15 and at 35 cm depth. From the total of 36 planar pH optodes sensors (18 at 15 cm and 18 at 35 cm depth) placed within the 18 rhizoboxes, maize roots crossed 12 and six of the sensors placed at 15 and 35 cm depth, respectively. The contact among roots and the planar pH optodes sensors can be reduced due to root curvature, especially for optodes placed in deeper depths [31]. In general, root tips reached the sensors at 15 cm depth at 4 DAG and the sensors at 35 cm depth at 8 DAG, except for BD + NPK L/L, where root tips reached the lower sensor at 11 DAG. Roots crossed the entire surface of the sensors within 24 h.

Bulk substrate pH over the experimental period is shown in **Table 6**. Considering that the pH of the arable field soil and of the peat substrate used to prepare the maize growing medium were 6.3 and 6.0, respectively, as a general trend, an acidification of the substrate was observed after the application of the treatments. Within C/L condition, bulk substrate pH values at 15 cm depth for BD, BD + NPK and NPK

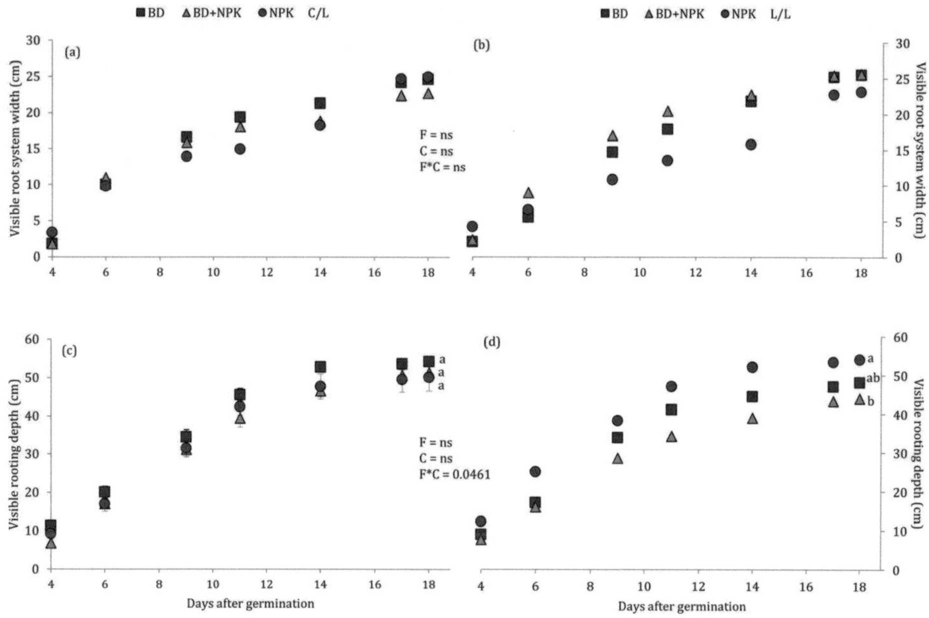

Figure 5.
*Effect of fertilizer type (Biogas Digestate - BD, Biogas Digestate+mineral NPK fertilizer - BD + NPK, mineral NPK fertilizer - NPK) and compaction condition (layered compacted substrate - C/L, loose substrate - L/L) on maize visible root system width (a, b) and visible rooting depth (c, d) over days after germination. Data are mean (n = 5) and error bars are standard error. Means with the same letter within compaction condition (C/L, L/L) at 18 days after germination are not statistically different according to Tukey's test at $p < 0.05$. F = effect of fertilizer; C = effect of compaction; F*C = interaction between fertilizer and compaction; ns = not significant.*

ranged from 4.3 to 5.2, 4.4 to 5.1, and 4.6 to 5.3, respectively, between 2 and 18 DAG (**Table 6**). At 35 cm depth, bulk substrate pH values ranged from 4.9 to 5.2, 4.6 to 5.2, and 4.8 to 5.4 for BD, BD + NPK and NPK, respectively (**Table 6**). Within L/L condition, bulk substrate pH values at 15 cm depth for BD, BD + NPK and NPK varied as follows: 4.6–5.2, 4.5–5.1, and 4.8–5.3, respectively (**Table 6**). Yet, in L/L condition, bulk substrate pH value ranges at 35 cm depth for BD, BD + NPK and NPK were 4.6–5.1, 4.9–5.3, and 4.9–5.3, respectively (**Table 6**). Differences between bulk substrate pH at 2 and 18 DAG within the same treatment were at maximum 0.4 pH units, which was observed for BD + NPK C/L at 15 cm (decrease of pH from 5.1 to 4.7) and at 35 cm (decrease of pH from 5.2 to 4.8) (**Table 6**).

Rhizosphere pH variation in relation to the bulk substrate over the experimental period is shown in **Figure 6**. Regarding the rhizosphere pH variation promoted by maize roots in relation to the bulk substrate pH, as a general trend, an alkalinization of rhizosphere pH was observed in all treatments at both 15 and 35 cm depth, except for BD + NPK 15 C/L and for BD 15 L/L, where an acidification of the rhizosphere pH up to −0.35 units was detected (**Figure 6**).

Under C/L condition, maximum rhizosphere pH alkalinization (+0.57 units) promoted by maize roots occurred in BD 15 at 8 DAG (**Figure 6a**). Similar alkalinization magnitude (+0.53 units) was also observed for BD 35 at 10 DAG (**Figure 6a**). Alkalinization promoted by NPK and BD + NPK application was in general less pronounced. The rhizosphere alkalinization was maintained up to maize harvest (18 DAG) with magnitudes of +0.43, +0.38, +0.23, and +0.16 pH units for BD 35, BD 15, NPK 15, and BD + NPK 35, respectively (**Figure 6a**). As previously mentioned,

DAG	BD 15	BD 35	BD + NPK 15	BD + NPK 35	NPK 15	NPK 35
			C/L			
2	5.0(0.0)	5.2(0.0)	5.1(0.1)	5.2(0.1)	4.7(0.1)	5.0(0.0)
4	4.3(0.1)	–[1]	4.7(0.2)	—	4.6(0.1)	—
5	4.7(0.0)	—	4.8(0.1)	—	4.9(0.1)	—
6	5.0(0.0)	—	5.0(0.1)	—	5.2(0.1)	—
8	5.0(0.1)	4.9(0.1)	4.8(0.2)	4.6(0.0)	5.1(0.1)	5.0(0.0)
9	5.2(0.0)	5.3(0.1)	5.0(0.2)	5.0(0.0)	5.2(0.1)	5.4(0.0)
10	4.6(0.0)	5.0(0.1)	4.4(0.2)	4.8(0.1)	4.8(0.1)	5.0(0.0)
11	4.9(0.1)	4.9(0.1)	4.7(0.1)	4.6(0.0)	5.0(0.1)	4.8(0.0)
12	5.2(0.0)	5.1(0.1)	5.0(0.2)	4.7(0.0)	5.3(0.1)	5.0(0.0)
13	4.9(0.0)	5.2(0.1)	4.8(0.1)	4.9(0.0)	5.0(0.1)	5.2(0.0)
17	5.1(0.0)	5.2(0.1)	5.0(0.1)	4.8(0.0)	5.3(0.1)	5.1(0.0)
18	5.0(0.1)	5.2(0.1)	4.7(0.2)	4.8(0.0)	4.9(0.1)	5.0(0.0)
			L/L			
2	5.1(0.1)	4.9(0.1)	5.0(0.1)	5.3(0.1)	4.8(0.1)	4.9(0.1)
4	4.7(0.1)	—	4.7(0.0)	—	4.7(0.0)	—
5	4.8(0.1)	—	4.9(0.0)	—	4.9(0.0)	—
6	5.0(0.1)	—	5.1(0.0)	—	5.2(0.0)	—
8	4.9(0.0)	4.7(0.1)	4.8(0.2)	5.0(0.1)	5.1(0.0)	5.0(0.1)
9	5.1(0.0)	5.1(0.0)	5.0(0.2)	5.3(0.1)	5.3(0.0)	5.3(0.0)
10	4.6(0.0)	4.8(0.0)	4.5(0.1)	4.9(0.1)	4.9(0.0)	5.0(0.1)
11	4.9(0.0)	4.6(0.1)	4.9(0.1)	4.9(0.1)	5.0(0.0)	4.9(0.1)
12	5.1(0.0)	4.9(0.0)	5.1(0.1)	5.1(0.1)	5.2(0.0)	5.1(0.0)
13	5.0(0.1)	5.0(0.1)	4.7(0.2)	5.2(0.0)	5.0(0.0)	5.2(0.0)
17	5.2(0.1)	5.0(0.1)	5.0(0.2)	5.1(0.1)	5.2(0.0)	5.1(0.1)
18	4.9(0.0)	4.9(0.0)	4.9(0.1)	5.1(0.0)	4.8(0.0)	5.0(0.1)

[1]*not determined.*

Table 6.
Bulk substrate pH over days after germination (DAG) of maize plants measured by optodes positioned at 15 and 35 cm depth in response to fertilizer type (Biogas Digestate - BD, Biogas Digestate+mineral NPK fertilizer - BD + NPK, mineral NPK fertilizer - NPK) and compaction condition (layered compacted substrate - C/L, loose substrate - L/L). Data are mean followed by standard error within parentheses.

within C/L condition rhizosphere pH acidification induced by roots occurred only in BD + NPK 15, where changes in rhizosphere pH oscillated between −0.35 (5 DAG) and +0.15 units (8 DAG) over the experimental period (**Figure 6a**).

Rhizosphere alkalization promoted by maize roots among fertilizers under L/L was remarkably distinct from that observed under C/L. Under L/L, the greatest rhizosphere alkalinization was observed in BD + NPK 15 at 9 DAG (+0.65) (**Figure 6b**). Such alkalinization magnitude was followed by NPK 15 at 5 DAG

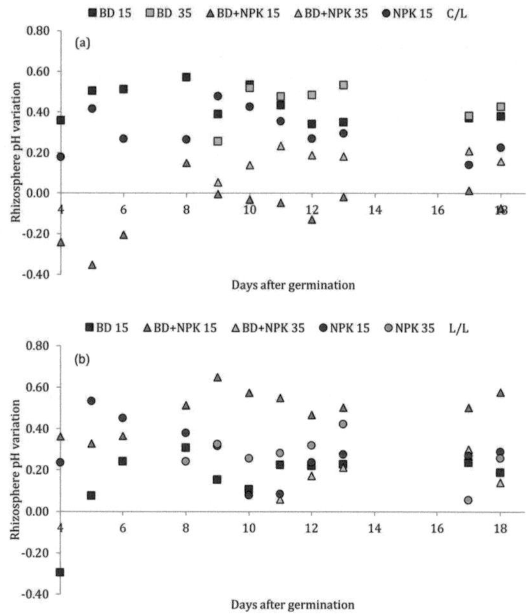

Figure 6.
Effect of fertilizer type (Biogas Digestate - BD, Biogas Digestate+mineral NPK fertilizer - BD + NPK, mineral NPK fertilizer - NPK) within layered compacted substrate (C/L) (a) and loose substrate (L/L) (b) condition on rhizosphere pH variation in relation to the bulk substrate pH. Rhizosphere and bulk substrate pH were measured by planar pH optodes positioned at 15 and 35 cm depth. Data are mean from different number of replicates, as it follows: under C/L - BD 15 (n = 3), BD 35 (n = 1), BD + NPK 15 (n = 1), BD + NPK 35 (n = 2), NPK 15 (n = 1), NPK 35 (n = 0); and, under L/L – BD 15 (n = 3), BD 35 (n = 0), BD + NPK 15 (n = 2), BD + NPK 35 (n = 1), NPK 15 (n = 2), NPK 35 (n = 2).

(+0.53) (**Figure 6b**), while a less pronounced alkalinization up to +0.31 units was observed in BD 15 at 8 DAG (**Figure 6b**). Rhizosphere alkalinization was predominant over acidification until maize harvest in all treatments and depths of pH measurement, when alkalinization magnitudes of +0.58, +0.29, +0.26, +0.19, and +0.14 units were observed in BD + NPK 15, NPK 15, NPK 35, BD 15 and BD + NPK 35, respectively (**Figure 6b**). In contrast to C/L condition, acidification of rhizosphere pH among treatments under L/L occurred only in BD 15 at 4 DAG (−0.29 units) (**Figure 6b**).

4. Discussion

4.1 Maize performance

Regarding substrate compaction condition, the low density of the peat substrate (0.21 g cm^{-3}) added to the field soil enabled the preparation of the loose layer in both C/L and L/L conditions. Maize roots subjected to C/L condition faced penetration resistance of 1.0 MPa and a dry bulk density of 0.84 g cm^{-3} (0–25 cm depth), which were 25- and 1.4-fold higher compared to L/L, respectively (0.04 MPa and 0.62 g cm^{-3}). It should be mentioned that substrate or soil bulk density and mechanical resistance may not have a linear relationship, which is in line with values obtained in rhizoboxes filled with loose and compacted substrate by Pfeifer et al. [11]. Yet, significant positive correlation between visible root length obtained with GROWSCREEN-Root and total root length of maize determined destructively, both under two levels of soil compaction (0.07 and 0.16 MPa) in a 4-week rhizobox experiment, has previously been reported by Nagel et al. [18]. These findings indicate that compaction does not affect the proportion of the roots visible at the transparent face of the rhizobox, at least for early maize root growth. Therefore, in our study we assume that compaction did not affect the proportion of maize roots visible at the transparent face of the rhizobox, and thus differences in root architecture are attributed to treatment effects.

Interestingly, significant changes in root architecture of juvenile maize reported by Nagel et al. [18] were observed in plants subjected to an even narrower range of substrate compaction (0.07 and 0.16 MPa) compared to that of our study (0.04 and 1.0 MPa). Our findings showed a reduction of the total root biomass of juvenile maize plants receiving BD compared to NPK and BD + NPK fertilization under C/L condition (**Figure 3**). Plant root:shoot ratio is usually sensitive to nutrient deficiency, water stress, high plant density, shading, and soil compaction [55–57]. As reported by Correa et al. [2], maize root:shoot ratio at 42 days after planting decreased significantly when substrate density was increased by 12%. In our study, maize root:shoot ratio was not affected by substrate compaction condition, but root:shoot ratio mean of fertilizer types (n = 10) differed statistically (NPK > BD) (**Table 3**). According to Bhattacharya [58], adequate P supply is known to enhance overall crop development and to promote early root formation and proliferation. Thus, the greater P supply via NPK fertilizer in comparison with BD (**Table 1**) may explain the higher root:shoot ratio (regardless of compaction condition, n = 10) (**Table 3**), shoot P content (regardless of compaction condition, n = 10) (**Table 4**), and total root dry weight (C/L) of maize plants receiving NPK instead of BD fertilization (**Figure 3**). These findings may be preferentially assigned to P instead of N plant nutrition since N content in maize shoots and roots promoted by BD and NPK fertilization did not differ from each other (**Table 4**).

Concerning root system architecture, Thaler and Pagès [55] reported that primary roots may be more sensitive than laterals to an increase in soil compaction. Yet, Bingham and Bengough [10] reported a decrease of 65% and 47% of barley main axis length and lateral root length, respectively, with an increase in soil compaction (from 0.3 to 1 MPa). In our study, under L/L condition (0.04 MPa), plants fertilized with BD + NPK and BD were able to develop roots deeper compared to NPK at harvest (**Figure 5d**), while higher visible main root length was obtained in NPK treatment (**Figure 4b**). The visible main roots length between treatments at 4 DAG, ranging from 10.0 to 13.1 cm (**Figure 4a** and **b**), indicates that maize germination and initial rooting were not negatively affected by BD fertilization (BD and BD + NPK) compared to NPK. This is in line with a maize germination rhizobox experiment that included two field soils mixed with peat and application of either maize silage-derived digestate or mineral NPK as fertilizer [49]. In relation to visible total lateral root length, in our study it was first determined at 6 DAG, and at harvest it was only affected by fertilizer type, where BD fertilization effect was comparable to that of NPK, regardless of compaction condition (see **Figure 4d** and its statistical footnote). The formation of lateral roots enables plants to adjust to heterogeneous situations in the soil, especially under favorable conditions of water or nutrient availability [11]. Thus, we may assume that the substrate fertility condition promoted by BD and NPK affected juvenile maize lateral root length in similar magnitude.

In general, N, P, and K content of maize shoots (**Table 4**) of all treatments indicated that plants were sufficiently supplied with these macronutrients [59]. Despite the sufficient nutrition status of maize shoots in response to the different fertilizer types and substrate compaction conditions, maize plants responded sensitively to the different treatments regarding nutrient content in shoots (P) and roots (N, P, and K). Sufficient N supply to maize plants via BD could be expected since the ordinary BD application doses of 40 Mg ha^{-1} contained an N amount equivalent to that provided by NPK (**Table 1**), which in turn was equivalent to an ordinary NPK fertilization for maize [49]. Similarly, sufficient P and K maize nutrition via BD was expected because the BD used in this experiment presented high P and K content (**Table 1**). This is attributed to the co-digestion of maize silage with chicken manure, once both residues are rich in K, and because chicken manure is particularly expected to also add significant amounts of P to the BD due to animal excretions [33]. In fact, the maize silage+chicken manure-derived BD used in our study contained similar N but higher P and K content compared to another BD tested by Robles-Aguilar et al. [49] for maize fertilization, where the BD used by the authors was produced under similar conditions compared to our study but had exclusively maize silage as feedstock. Still with respect to sufficient N supply to maize via BD fertilization, two aspects should be mentioned. First, BD contains great amounts of readily available N forms (mainly NH_4^+). Second, the BD tested in our study presented a low C:N ratio of 6 that may have facilitated its fast degradation by the microbiota, consequently leading to a flush of available N in the growing medium, as previously reported by Andruschkewitsch et al. [40]. In this sense, we consider that BD mineralization, BD-N-released forms, and their transformation in soils should be addressed in further studies to enhance our understanding regarding nutrient cycling and timing between BD fertilization and crop sowing in BD-amended soils.

The SPAD readings were similar across all treatments (**Table 3**), indicating similar N leaf status of maize plants [60], and corroborating the similar shoot N content verified for the different treatments (**Table 4**). The SPAD readings observed in our study suggest that there was no N starvation during maize juvenile development (18 DAG) [61]. Furthermore, the SPAD values reported in our study were higher than that reported by Robles-Aguilar et al. [49] for maize plants (from 25 to 45 DAG) fertilized with 40 Mg ha^{-1} of pure maize silage-derived digestate. The dose of N application via BD (212 kg N ha^{-1}) along with the fact that N in the BD composition is predominantly in the form of NH_4^+, may explain the sufficient N nutrition of maize in our study. According to Lukehurst et al. [33], during anaerobic digestion part of the N in organic form is converted to NH_4^+, and this is the main component readily available to plants after fertilization with BD. In fact, substrate N content at 25–55 cm depth (regardless of compaction condition, n = 10) was higher in BD than in NPK (**Table 5**). Linking such results with the fact that maize shoot and root N content were not statistically different between BD and NPK, and that N amendments were similar between fertilizer type (**Table 1**), it may be suggested that N losses from the total substrate profile (0–55 cm) were possibly attenuated by the presence of BD. This hypothetical prevention of N leaching by BD may be tested in future studies.

Shoot K content of maize plants fertilized with BD + NPK was equivalent to that fertilized with NPK (regardless of compaction condition, n = 10), as well as root K content under L/L (**Table 4**). Here it is important to highlight that although the amount of K added by BD to the substrate was 154% higher than that of NPK fertilizer, BD did not cause an excess of K content in shoot tissue in relation to NPK. Possibly, part of the K applied via BD may have been leached during the period between the experiment's beginning and soil sampling at maize harvest. In our study, nutrient losses via leaching were not determined. In this sense, we consider leachate evaluation to be addressed in further studies to enhance our knowledge on closing the nutrient cycles in BD-amended soils. Our findings show the importance of supporting and planning the use of BD derived from maize silage+chicken manure in cropping systems. This type of BD, besides sufficiently supplying maize seedlings with N, P, and K, as verified in our study, may contribute to, at least partially, covering maize's increasing K demands in the next development stages, especially if K leaching can be in parallel attenuated.

Interestingly, under C/L condition, root K content was higher in plants fertilized with NPK than with BD and BD + NPK, following an inverse order of amount of K amendment via fertilization (BD > BD + NPK > NPK) (**Table 1**). At 0–25 cm depth, substrate K content was lower under C/L than under L/L condition (regardless of fertilizer type, n = 15). Thus, the top compacted layer in C/L condition may have partially and temporarily reduced the downward movement of K, and subsequently, it may have enabled greater K uptake by plants in the main rooting zone (0–25 cm depth), where its uptake is facilitated due to the greater concentration of roots compared to 25–55 cm depth (**Figure 3**). This is reinforced by the fact that NPK fertilization resulted in lower substrate K content at 0–25 cm depth (regardless of compaction condition, n = 10) and in higher root K content in comparison to BD and BD + NPK. In addition, the effect of compaction on preventing K leaching presumably had particular importance in the substrate with expected lower cation exchange capacity (NPK). In this view, BD fertilization is expected to increase substrate C content cation exchange capacity due to the addition of organic functional groups commonly present in organic residues. Soil cation exchange capacity increase in response to BD application was reported by Glowaca et al. [62]. In our study, increased substrate cation exchange capacity due to BD application may have occurred preferentially at 25–55 cm depth, where substrate C content was higher in BD fertilization compared to NPK (**Table 5**). Yet, the greater substrate C content in BD compared to NPK observed only at 25–55 cm depth may be partially explained by the possible leaching of soluble organic compounds commonly present in digestates [33, 41].

Under L/L condition, a top compacted layer that may partially reduce K leaching is absent, so K leaching is expected to increase (BD < BD + NPK < NPK) as lower is the assumed capacity of the substrate to adsorb it (NPK < BD + NPK < BD). This hypothesis may be confirmed in complementary studies, but this interpretation is corroborated by the observation that root K content under L/L was higher in BD than in NPK, whereas BD + NPK did not differ from the other treatments (**Table 4**). In this view, root K content was greater as higher was the presumed capacity of the substrate to retain cations, and it also seemingly followed the amount of K application via fertilizers (**Table 1**). Also, at 25–55 cm depth, where levels of substrate compaction were similar among C/L and L/L condition (dry bulk density of 0.64 and 0.61 g cm^{-3}, respectively), greater substrate K content was obtained in NPK and BD + NPK treatments compared to BD (**Table 5**). These findings may indicate greater K leaching in treatments receiving NPK (and less in BD). Accordingly, Pfeifer et al. [11] reported greater downward movement of K in loose than in compacted substrate in rhizoboxes receiving similar fertilization and watering. The authors attributed the lower K leaching in the compacted substrate to the physical impedance for downward water flow. The levels of substrate compaction (0.04–1.23 MPa) applied by Pfeifer et al. [11] were comparable to that of our study (0.04–1.00 MPa), giving support to our interpretation. The significant effect of interaction between fertilizer type and compaction condition on root K content (**Table 2**) is of particular interest once field soils may be subjected to some compaction level during the crop season.

Several studies have reported the benefits of digestates for plant growth and nutrition [36, 37, 39, 63]. However, when digestates do not comply with crop demand for N, P, and K fertilization, then mineral fertilizer supplementation is required. In this case, special attention has to be given to the nutrient balance of fertilizers. Taking into account the N:P:K ratios of BD (1.00:0.26:1.28) and NPK (1.00:0.33:0.53), it seems that the BD + NPK mixture resulted in an intermediate N:P:K ratio (1.00:0.30:0.92) for both P and K and led to plant performance similar to NPK. Regarding P supply, despite the lower P content in the shoots and roots of juvenile maize plants fertilized with BD compared to those fertilized with NPK (regardless of compaction condition, n = 10), it did not affect maize shoot dry weight (**Figure 3**), number of leaves, SLA, and SPAD readings (**Table 3**). However, the lower P content in maize plants fertilized with BD may explain its lower root:shoot ratio in comparison with NPK (**Table 3**). While, BD + NPK promoted similar maize shoot P content and root:shoot ratio to that obtained with NPK, without negative implications to overall shoot parameters. Yet, root system development in terms of visible length of main roots and total lateral roots, as well as root system width and rooting depth of maize plants fertilized with BD + NPK was comparable to that of NPK. These findings reinforce the potential of BD fertilization, possibly in combination with NPK, to reduce environmental liabilities associated with mineral fertilizer utilization and improper BD disposal in the environment, without compromising initial maize development. These findings may be corroborated by studies covering the complete maize growing cycle.

4.2 Rhizosphere and bulk substrate pH over days after germination

The acidic pH of the substrate after the application of the fertilizers (BD, BD + NPK and NPK) (**Table 6**) in relation to the pH of the field soil (6.3), peat (6.0) and BD (8.2), is probably associated with a combination of factors, likely: (i) fast ammonium nitrification (NH_4^+ + $2O_2$ NO_3^- + H_2O + $2H^+$), since about 60, 61 and 63% of the total N of BD, BD + NPK and NPK was in the form of NH_4^+, respectively, and drainage holes at the bottom of the rhizoboxes allowed oxygenation of substrate [64, 65]; (ii) reaction between Mg^{2+}, NH_4^+ and PO_4^{3-} ions resulting in a release of H^+ ions to the solution [65]; (iii) partial and temporal occupation of substrate cation exchange sites by NH_4^+ instead of by basic cations as Ca^{2+}, Mg^{2+} and K^+; and, (iv) leaching of basic cations as Ca^{2+}, Mg^{2+}, Na^+ and K^+, especially due to the low cation exchange capacity of the substrates (mainly the one without BD application) and its replacement with H^+. The H^+ in this case is also expectedly derived from the fast mineralization of the organic materials in the substrate, e.g., soil organic matter from the field soil used and BD, due to their low C:N ratio, 11 and 6, respectively.

Organic matter mineralization is intensified under these circumstances due to the fast and high input of available N in the substrates (derived from BD and NPK) and substrate wetting, stimulating microbial activity, and because BD usually contains easily biodegradable organic compounds [66]. Here, it is important to mention that about 4 to 5 days elapsed between the substrate's preparation with fertilizers and the first bulk substrate pH measurements with planar optodes (2 DAG). During this time, the installation of the planar pH optode sensors into rhizoboxes was performed, the 30 rhizoboxes were prepared with the substrates and levels of compaction, and maize was sown. This has been shown to be sufficient time to trigger a significant part of the substrate acidification mechanisms aforementioned [40, 49, 65–68], also due to substrate watering performed to stimulate seed germination. Rhizosphere pH measurements started at 4 DAG, when the first maize roots reached the planar optode sensors positioned at 15 cm depth.

The substrate acidification and NH_4^+ nitrification seem to have strongly driven the rhizosphere pH changes induced by maize roots observed in our study. Both factors settled the acidic bulk substrate pH in which maize roots were growing (**Table 6**), and due to nitrification, NO_3^- possibly became the main N source to plants over DAG instead of NH_4^+. Additionally, NO_3^- was also directly added to the substrates via both NPK and BD fertilizers (see Section 2.1). This may explain the overall trend of rhizosphere pH alkalinization induced by maize roots at both 15 and 55 cm depth during the experimental period. Slight and exceptional rhizosphere acidification was observed in a few measurements (**Figure 6**). According to Custos et al. [24], in aerobic soils, NO_3^- was the most absorbed nutrient (compared to K, NH_4^+, Ca, Mg, SO_4^{2-}, and P ions) by maize plants within the 1 and 4 weeks of growth, resulting in a strong depletion of anions (especially NO_3^-) and ultimately in a positive charge balance and alkalinization of rhizosphere pH. The authors estimated that absorption of NO_3^- was 837 and 830% greater than that of NH_4^+ within the 1 and 4 weeks of maize growth, respectively. The authors also stated that rhizosphere alkalinization by maize roots was associated with NO_3^- depletion in the rhizosphere and that NO_3^- uptake was equilibrated by H^+ absorption by maize roots with equivalent excretion of OH^-. Conversely, in the case of NH_4^+ uptake, plants tend to maintain the electro-neutrality across the root membrane by releasing H^+, which would consequently lead to rhizosphere acidification [22, 69]. In this way, plant N uptake as NH_4^+ or as NO_3^- usually results in rhizosphere acidification or alkalinization, respectively [29, 70–74], and such plant N uptake mechanisms have been reported for maize and other species such as ryegrass, rice, alpine pennycress, and wheat [24, 31, 52, 75, 76].

Faget et al. [77] found similar temporal and magnitude of rhizosphere alkalinization induced by maize roots (measured with planar pH optodes) compared to our study. The authors observed maximum alkalinization of up to 0.62 pH units in relation to the substrate at 8 days after maize transplantation to rhizoboxes. The authors attributed their results to the fact that NO_3^- was the only source of N for maize plants. Similarly to our study, the authors found that alkalinization persisted when coming closer to harvest (14 days after transplanting), but in a lower magnitude compared to our data.

Our data showed that when maize plants were supplied with less P (without P deprivation) and more N, as in the case of BD compared to BD + NPK and NPK fertilizers, roots still tended to alkalinize the rhizosphere (**Figure 6**). Li et al. [78] observed that under comparable sufficient N and deficient P supply, maize roots alkalinized the rhizosphere, while faba bean acidified it. According to the authors, faba bean roots mobilized P by releasing organic acids and protons to the rhizosphere, which also favored P uptake by maize when both species were intercropped. These findings reinforce that mobilization of P by maize roots is unlikely to occur. Therefore, the type of BD used in this experiment should be ideally complemented with high-soluble P sources, such as other organic materials or mineral P, in order to avoid possible P deficiency in maize plants. Also, maize could be intercropped with legume plants able to mobilize P. In this view, rhizosphere alkalinization promoted by maize roots under monoculture and intercropping was reported by Zhou et al. [79]. Furthermore, these authors observed that alkalinization by maize roots was higher under low P supply, very similarly to what was observed at 15 cm depth for plants of BD treatment in our study (**Figure 6**). At 35 cm depth, such observation was not as clear as it was for the upper depth, but still the highest values of rhizosphere alkalinization at 35 cm depth were verified in a BD fertilization treatment (BD + NPK) and not in NPK alone (**Figure 6**), which supplied the highest P dose to maize (**Table 1**).

In this context, the general pattern of rhizosphere alkalinization observed for most treatments over the experimental period, as well as the different magnitude of such alkalinization between treatments, may be closely related to the dynamics of NH_4^+ nitrification, corroborating previous findings of rhizosphere pH change induced by maize roots [78, 79]. Overall, our rhizosphere pH investigation indicates that although the main form of N in the BD was NH_4^+, the N uptake by maize in BD treatments may have occurred predominantly as NO_3^-. The short time between BD amendment, nitrification, and NO_3^- uptake by maize should be taken into account when BD fertilization and maize sowing are planned in order to further close N cycle in BD-amended soils. Therefore, specific BD mineralization and N speciation studies should be performed, and microbiological parameters should be addressed to improve our understanding of the dynamics of N uptake by early growth of crops fertilized with BD.

In our study, the acid-growth mechanism, characterized by a local restricted and preferential cation uptake (mainly K^+ and NH_4^+) in maize root sections [31], was not evidenced by our data, although this was not the focus of our investigation.

5. Conclusion

Efficient maize fertilization with BD from the co-digestion of maize silage and chicken manure offers a great opportunity to improve nutrient cycling in bioenergy-crops production. Negative effects of BD on maize germination were not observed in this study. Our findings indicated that both NPK and BD fertilizers, alone or combined, supplied maize sufficiently with N, P, and K, and plants had similar specific leaf area and chlorophyll content. Nevertheless, plants fertilized solely with BD had lower shoot and root P content compared to NPK, irrespective of compaction condition. Yet, compaction led to lower shoot and root P contents compared to loose substrate. This may be associated with the fact that P in BD may not be readily available to plants as in NPK. In response to lower shoot and root P content, plants fertilized with BD had reduced root biomass (0–55 cm) (C/L) and main root length (L/L) in relation to NPK, and apparently plants fertilized with BD compensated for it by growing deeper under loose substrate in relation to NPK. The lower root:shoot ratio of plants receiving BD fertilization in relation to NPK was attributed to lower maize P uptake in BD.

Overall, maize roots induced rhizosphere alkalinization, regardless of treatment and substrate depth. This was attributed to preferential NO_3^- uptake, even though most of N in the fertilizers was initially in the form of NH_4^+. This was assigned to the possible fast nitrification of NH_4^+. Magnitude of rhizosphere alkalinization between treatments varied, and it was attributed to possible specific dynamics of nitrification in each treatment. To address such hypotheses, studies dedicated to investigating mineralization of BD and nitrification depending on fertilizer type are needed to help find time-wise BD amendment and crop sowing. Maize plants supplied with less P (not P deprivation) did not induce rhizosphere acidification to mobilize P, as reported in the literature for some legume species.

In general, we confirmed our hypothesis that maize shoot and root seedling parameters under mineral NPK fertilization alone or mixed with BD would be equivalent. Exceptions occurred only for root P content and visible rooting depth in loose substrate, which were slightly higher in NPK fertilization used alone. Our findings suggest that maize fertilization with the type of BD tested in our study (maize silage+chicken manure) may be combined with fast-soluble P sources, or maize should be intercropped with P-mobilizing plant species to possibly deliver maize performance equivalent to regular NPK fertilization. These hypotheses may be tested in future studies, also possibly covering more maize development stages, aiming for further insights for optimization of BD as fertilizer.

Acknowledgements

Thanks to Prof. Dr. Vicky Temperton and Dr. Wilfried Wolff for their help during data preparation. We thank the Central Institute for Engineering, Electronics and Analytics (ZEA) – Forschungszentrum Jülich for the analytical measurements. We are grateful to ADRW Naturpower GmbH & Co. Kg for providing the biogas digestate and PreSens GmbH for all technical support. Dr. Daniela P.B. Leal received funding from the International Jülich PostDoc Programme within the joint project "BioEconomy and plant sciences: Root system phenotypes for sustainable soil exploitation in conservation land management." Further funding for the scientific investigation was provided by the German Federal Ministry of Education and Research (BMBF) within PURESBio project (grant number 031A289A).

Conflict of interest

The authors declare no conflict of interest.

Author details

Daniela Piaz Barbosa Leal[1], Otávio dos Anjos Leal[2*], Stephan Blossfeld[1,3] and Nicolai David Jablonowski[1]

1 Institute of Bio- and Geosciences, IBG-2: Plant Sciences, Forschungszentrum Jülich GmbH, Jülich, Germany

2 Institute of Bio- and Geosciences, IBG-3: Agrosphere, Forschungszentrum Jülich GmbH, Jülich, Germany

3 Gesamtschule Kürten, Kürten, Germany

*Address all correspondence to: o.dos.anjos.leal@fz-juelich.de

References

[1] Lipiec J, Horn R, Pietrusiewicz J, Siczek A. Effects of soil compaction on root elongation and anatomy of different cereal plant species. Soil and Tillage Research. 2012;**121**:74-81. DOI: 10.1016/j.still.2012.01.013

[2] Correa J, Postma JA, Watt M, Wojciechowski T. Soil compaction and the architectural plasticity of root systems. Journal of Experimental Botany. 2019;**70**:6019-6034. DOI: 10.1093/jxb/erz383

[3] Clark LJ, Whalley WR, Barraclough PB. How do roots penetrate strong soil? Plant and Soil. 2003;**255**:93-104. DOI: 10.1023/a:1026140122848

[4] Grzesiak S, Grzesiak MT, Hura T, Marcińska I, Rzepka A. Changes in root system structure, leaf water potential and gas exchange of maize and triticale seedlings affected by soil compaction. Environmental and Experimental Botany. 2013;**88**:2-10. DOI: 10.1016/j.envexpbot.2012.01.010

[5] Hamza MA, Anderson WK. Soil compaction in cropping systems. A review of the nature, causes and possible solutions. Soil and Tillage Research. 2005;**82**:121-145. DOI: 10.1016/j.still.2004.08.009

[6] Kaufmann M, Tobias S, Schulin R. Comparison of critical limits for crop plant growth based on different indicators for the state of soil compaction. Journal of Plant Nutrition and Soil Science. 2010;**173**:573-583. DOI: 10.1002/jpln.200900129

[7] Bengough AG, McKenzie BM, Hallett PD, Valentine TA. Root elongation, water stress, and mechanical impedance: A review of limiting stresses and beneficial root tip traits. Journal of Experimental Botany. 2011;**62**:59-68. DOI: 10.1093/jxb/erq350

[8] Nosalewicz A, Lipiec J. The effect of compacted soil layers on vertical root distribution and water uptake by wheat. Plant and Soil. 2014;**375**:229-240. DOI: 10.1007/s11104-013-1961-0

[9] Montagu KD, Conroy JP, Atwell BJ. The position of localized soil compaction determines root and subsequent shoot growth responses. Journal of Experimental Botany. 2001;**52**:2127-2133. DOI: 10.1093/jexbot/52.364.2127

[10] Bingham IJ, Bengough AG. Morphological plasticity of wheat and barley roots in response to spatial variation in soil strength. Plant and Soil. 2003;**250**:273-282. DOI: 10.1023/a:1022891519039

[11] Pfeifer J, Faget M, Walter A, Blossfeld S, Fiorani F, Schurr U, et al. Spring barley shows dynamic compensatory root and shoot growth responses when exposed to localised soil compaction and fertilization. Functional Plant Biology. 2014;**41**:581-597. DOI: 10.1071/FP13224

[12] Popova L, Dusschoten D, Nagel KA, Fiorani F, Mazzolai B. Plant root tortuosity: An indicator of root path formation in soil with different composition and density. Annals of Botany-London. 2016;**118**:685-698. DOI: 10.1093/aob/mcw057

[13] Calogeno JC, Raphael JPA, Rigon JPG, Oliveira Neto L, Rosolem CA. Soil compaction management and soybean yields with cover crops under no-till and occasional chiseling. European Journal of Agronomy.

2017;**85**:31-37. DOI: 10.1016/j.eja.2017.02.001

[14] Kassam A, Friedrich T, Derpsch R. Global spread of conservation agriculture. International Journal of Environmental Studies. 2018;**787**:1-23. DOI: 10.1080/00207233.2018.1494927

[15] Pott LP, Amado TJC, Leal OA, Ciampitti IA. Mitigation of soil compaction for boosting crop productivity at varying yield environments in southern Brazil. European Journal of Soil Science. 2019;**71**:1-16. Available from: https://bsssjournals.onlinelibrary.wiley.com/doi/abs/10.1111/ejss.12880. DOI: 10.1111/ejss.12880

[16] Canarache A. PENETR – A generalized semi-empirical model estimating soil resistance to penetration. Soil and Tillage Research. 1990;**16**:51-70. DOI: 10.1016/0167-1987(90)90021-5

[17] Moraes MT, Silva VR, Zwirtes AL, Carlesso R. Use of penetrometers in agriculture: A review. Engenharia Agrícola. 2014;**34**:179-193. DOI: 10.1590/S0100-69162014000100019

[18] Nagel KA, Putz A, Gilmer F, Heinz K, Fischbach A, Pfeifer J, et al. GROWSCREEN-Rhizo is a novel phenotyping robot enabling simultaneous measurements of root and shoot growth for plants grown in soil-filled rhizotrons. Functional Plant Biology. 2012;**39**:891-904. DOI: 10.1071/FP12023

[19] Colombi T, Walter A. Root responses of triticale and soybean to soil compaction in the field are reproducible under controlled conditions. Functional Plant Biology. 2016;**43**:114-128. DOI: 10.1071/FP15194

[20] Passioura JB. Soil conditions and plant growth. Plant, Cell and Environment. 2002;**25**:311-318. DOI: 10.1046/j.0016-8025.2001.00802.x

[21] Haling RE, Brown LK, Bengough AG, Young IM, Hallett PD, White PJ, et al. Root hairs improve root penetration, root–soil contact, and phosphorus acquisition in soils of different strength. Journal of Experimental Botany. 2013;**64**:3711-3721. DOI: 10.1093/jxb/ert200

[22] Hinsinger P, Plassard C, Tang C, Jaillard B. Origins of root-mediated pH changes in the rhizosphere and their responses to environmental constraints: A review. Plant and Soil. 2003;**248**:43-59. DOI: 10.1023/a:1022371130939

[23] Haynes RJ. Active ion uptake and maintenance of cation-anion balance: A critical examination of their role in regulating rhizosphere pH. Plant and Soil. 1990;**126**:247-264. DOI: 10.1007/BF00012828

[24] Custos JM, Moyne C, Sterckeman T. How root nutrient uptake affects rhizosphere pH: A modelling study. Geoderma. 2020;**369**:1-12. DOI: 10.1016/j.geoderma.2020.114314

[25] Neumann G, George TS, Plassard C. Strategies and methods for studying the rhizosphere – The plant science toolbox. Plant and Soil. 2009;**321**:431-456. DOI: 10.1007/s11104-009-9953-9

[26] Blossfeld S, Gansert D. A novel non-invasive optical method for quantitative visualization of pH dynamics in the rhizosphere of plants. Plant, Cell and Environment. 2007;**30**:176-186. DOI: 10.1111/j.1365-3040.2006.01616.x

[27] Blossfeld S. Light for the dark side of plant life: Planar optodes visualizing rhizosphere processes. Plant and Soil. 2013;**369**:29-32. DOI: 10.1007/s11104-013-1767-0

[28] Koop-Jakobsen K, Mueller P, Meier RJ, Liebsch G, Jensen K. Plant-sediment interactions in salt marshes – An optode imaging study of O_2, pH, and CO_2 gradients in the rhizosphere. Frontiers in Plant Science. 2018;**9**:1-11. DOI: 10.3389/fpls.2018.00541

[29] Taylor AR, Bloom AJ. Ammonium, nitrate, and proton fluxes along the maize root. Plant, Cell and Environment. 1998;**21**:1255-1263. DOI: 10.1046/j.1365-3040.1998.00357.x

[30] Fan L, Neumann PM. The spatially variable inhibition by water deficit of maize root growth correlates with altered profiles of proton flux and cell wall pH. Plant Physiology. 2004;**135**:2291-2300. DOI: 10.1104/pp.104.041426

[31] Blossfeld S, Perriguey J, Sterckeman T, Morel JL, Loesch R. Rhizosphere pH dynamics in trace-metal-contaminated soils, monitored with planar pH optodes. Plant and Soil. 2010;**330**:173-184. DOI: 10.1007/s11104-009-0190-z

[32] Herrmann A. Biogas production from maize: Current state, challenges and prospects. 2. Agronomic and environmental aspects. Bioenergy Research. 2013;**6**:372-387. DOI: 10.1007/s12155-012-9227-x

[33] Lukehurst CT, Frost P, Al Seadi T. Utilisation of Digestate from Biogas Plants as Biofertilizer. Rotorua, New Zealand: IEA Bioenergy [Internet]; 2010. Available from: https://www.ieabioenergy.com/blog/publications/utilisation-of-digestate-from-biogas-plants-as-biofertiliser/ [Accessed: July 05, 2024]

[34] Gissén C, Prade T, Kreuger E, Nges IA, Rosenqvist H, Svensson S, et al. Comparing energy crops for biogas production-yields, energy input and costs in cultivation using digestate and mineral fertilization. Biomass and Bioenergy. 2014;**64**:199-210. DOI: 10.1016/j.biombioe.2014.03.061

[35] Wilken D, Rauh S, Fruhner-Weiß R, Strippel F, Bontempo G, Kramer A, et al. Digestate as Fertilizer. Freising: Fachverband Biogas e.V. [Internet]; 2018. Available from: https://www.biogas.org/fileadmin/redaktion/dokumente/medien/broschueren/digestate/digestate_as_fertilizer.pdf [Accessed: July 03, 2024]

[36] Möller K, Müller T. Effects of anaerobic digestion on digestate nutrient availability and crop growth: A review. Engineering in Life Sciences. 2012;**12**:242-257. DOI: 10.1002/elsc.201100085

[37] Tambone F, Scaglia B, D'Imporzano G, Schievano A, Orzi V, Salati S, et al. Assessing amendment and fertilizing properties of digestates from anaerobic digestion through a comparative study with digested sludge and compost. Chemosphere. 2010;**81**:577-583. DOI: 10.1016/j.chemosphere.2010.08.034

[38] Abubaker J, Risberg K, Pell M. Biogas residues as fertilizers – Effects on wheat growth and soil microbial activities. Applied Energy. 2012;**99**:126-134. DOI: 10.1016/j.apenergy.2012.04.050

[39] Walsh JJ, Jones DL, Edwards-Jones G, Williams AP. Replacing inorganic fertilizer with anaerobic digestate may maintain agricultural productivity at less environmental cost. Journal of Plant Nutrition and Soil Science. 2012;**175**:840-845. DOI: 10.1002/jpln.201200214

[40] Andruschkewitsch M, Wachendorf C, Wachendorf M. Effects of digestates from different biogas production systems on above and belowground grass growth and the nitrogen status of the plant-soil-system. Grassland Science. 2013;**59**:183-195. DOI: 10.1111/grs.12028

[41] Nkoa R. Agricultural benefits and environmental risks of soil fertilization with anaerobic digestates: A review. Agronomy for Sustainable Development. 2014;**34**:473-492. DOI: 10.1007/s13593-013-0196-z

[42] Arthurson V. Closing the global energy and nutrient cycles through application of biogas residue to agricultural land – Potential benefits and drawbacks. Energies. 2009;**2**:226-242. DOI: 10.3390/en20200226

[43] Theuerl S, Herrmann C, Heiermann M, Grundmann P, Landwehr N, Kreidenweis U, et al. The future agricultural biogas plant in Germany: A vision. Energies. 2019;**12**:396. DOI: 10.3390/en12030396

[44] Calderón C, Colla M, Jossart JM, Hemeleers N, Martin A, Aveni N, et al. Bioenergy Europe statistical report 2019. In: Report Pellet: Bioenergy Europe [Internet]. Brussels: Bioenergy Europe; 2019. Available from: https://biom.cz/upload/9982d8381d3da848a8072e06cf96ec87/sr19_biomass-for-heat_final-web.pdf [Accessed: July 05, 2024]

[45] Corden C, Bougas K, Cunningham E, Tyrer D, Kreissig J, Zettl E, et al. Digestate and Compost as Fertilizers: Risk Assessment and Risk Management Options. London: Wood [Internet]; 2019. Available from: https://ec.europa.eu/environment/chemicals/reach/pdf/40039%20Digestate%20and%20Compost%20RMOA%20-%20Final%20report%20i2_20190208.pdf [Accessed: July 04, 2024]

[46] Nabel M, Schrey SD, Poorter H, Koller R, Jablonowski ND. Effects of digestate fertilization on *Sida hermaphrodita*: Boosting biomass yields on marginal soils by increasing soil fertility. Biomass and Bioenergy. 2017;**107**:207-213. DOI: 10.1016/j.biombioe.2017.10.009

[47] Barbosa DBP, Nabel M, Jablonowski ND. Biogas-digestate as nutrient source for biomass production of *Sida hermaphrodita*, *Zea mays* L. and Medicago sativa L. Energy Procedia. 2014;**59**:120-126. DOI: 10.1016/j.egypro.2014.10.357

[48] Nabel M, Schrey SD, Poorter H, Koller R, Nagel KA, Temperton VM, et al. Coming late for dinner: Localized digestate depot fertilization for extensive cultivation of marginal soil with *Sida hermaphrodita*. Frontiers in Plant Science. 2018;**9**:1-14. DOI: 10.3389/fpls.2018.01095

[49] Robles-Aguilar AA, Temperton VM, Jablonowski ND. Maize silage digestate application affecting germination and early growth of maize modulated by soil type. Agronomy. 2019;**9**:473. DOI: 10.3390/agronomy9080473

[50] Dietrich CC, Rahaman MA, Robles-Aguilar AA, Latif S, Intani K, Müller J, et al. Nutrient loaded biochar doubled biomass production in juvenile maize plants (*Zea mays* L.). Agronomy. 2020;**10**:567. DOI: 10.3390/agronomy10040567

[51] Nagel KA, Bonnett D, Furbank R, Walter A, Schurr U, Watt M. Simultaneous effects of leaf irradiance and soil moisture on growth and root system architecture of novel wheat genotypes: Implications for phenotyping. Journal of Experimental Botany. 2015;**66**:5441-5452. DOI: 10.1093/jxb/erv290

[52] Blossfeld S, Schreiber CM, Liebsch G, Kuhn AJ, Hinsinger P. Quantitative imaging of rhizosphere pH and CO_2 dynamics with planar optodes. Annals of Botany-London. 2013;**112**:267-276. DOI: 10.1093/aob/mct047

[53] Gransee A, Wittenmayer L. Qualitative and quantitative analysis of water-soluble root exudates in relation to plant species and development. Journal of Plant Nutrition and Soil Science. 2000;**163**:381-385. DOI: 10.1002/1522-2624(200008)163:4<381::AID-JPLN381>3.0.CO;2-7

[54] Larsen M, Borisov SM, Grunwald B, Klimant I, Glud RN. A simple and inexpensive high resolution color ratiometric planar optode imaging approach: Application to oxygen and pH sensing. Limnology and Oceanography Methods. 2011;**9**:348-360. DOI: 10.4319/lom.2011.9.348

[55] Thaler P, Pagès L. Why are laterals less affected than main axes by homogeneous unfavourable physical conditions? A model-based hypothesis. Plant and Soil. 1999;**217**:151-157. DOI: 10.1023/a:1004677128533

[56] Amos B, Walters DT. Maize root biomass and net rhizodeposited carbon. Soil Science Society of America Journal. 2006;**70**:1489-1503. DOI: 10.2136/sssaj2005.0216

[57] Yu P, Li X, White PJ, Li C. A large and deep root system underlies high nitrogen-use efficiency in maize production. PLoS One. 2015;**10**:5. DOI: 10.1371/journal.pone.0126293

[58] Bhattacharya A. Changing environmental condition and phosphorus-use efficiency in plants. In: Bhattacharya A, editor. Changing Climate and Resource Use Efficiency in Plants. 1st ed. Oxford: Elsevier Ltd.; 2019. pp. 241-305. DOI: 10.1016/B978-0-12-816209-5.00005-2

[59] Campbell CR, Plank CO. Reference Sufficiency Ranges for Plant Analysis in the Southern Region of the United States. Raleigh: Southern Cooperative Series Bulletin. [Internet]; 2000. Available from: http://www.ncagr.gov/agronomi/saaesd/scsb394.pdf [Accessed: July 04, 2024]

[60] Xiong D, Chen J, Yu T, Gao W, Ling X, Li Y, et al. SPAD-based leaf nitrogen estimation is impacted by environmental factors and crop leaf characteristics. Scientific Reports. 2015;**5**:1-12. DOI: 10.1038/srep13389

[61] Argenta G, Silva PRF, Sangoi L. Leaf relative chlorophyll content as an indicator parameter to predict nitrogen fertilization in maize. Ciência Rural. 2004;**34**:1379-1387. DOI: 10.1590/S0103-84782004000500009

[62] Glowaca A, Szostak B, Klebaniuk R. Effect of biogas digestate and mineral fertilization on the soil properties and yield and nutritional value of switchgrass forage. Agronomy. 2020;**10**:490. DOI: 10.3390/agronomy10040490

[63] Gutser R, Ebertseder T, Weber A, Schraml M, Schmidhalter U. Short-term and residual availability of nitrogen after long-term application of organic fertilizers on arable land. Journal of Plant Nutrition and Soil Science. 2005;**168**:439-446. DOI: 10.1002/jpln.200520510

[64] Bernal MP, Kirchmann H. Carbon and nitrogen mineralization and ammonia volatilization from fresh, aerobically and anaerobically treated pig manure during incubation with soil. Biology and Fertility of Soils. 1992;**13**:135-141. DOI: 10.1007/BF00336268

[65] Le Corre KS, Valsami-Jones E, Hobbs P, Parsons SA. Phosphorus recovery from wastewater by struvite crystallization: A review. Critical Reviews in Environmental Science and Technology. 2009;**39**:433-477. DOI: 10.1080/10643380701640573

[66] Kamprath EJ, Smyth TJ. Liming. In: Hillel D, Hatfield JH, Powlson DS, Rosenzweig C, Scow KM, Singer MJ, Sparks DL, editors. Encyclopedia of Soils in the Environment. 1st ed. Oxford: Elsevier Ltd; 2005. pp. 350-358

[67] Möller K, Stinner W, Deuker A, Leithold G. Effects of different manuring systems with and without biogas digestion on nitrogen cycle and crop yield in mixed organic dairy farming systems. Nutrient Cycling in Agroecosystems. 2008;**82**:209-232. DOI: 10.1007/s10705-008-9196-9

[68] Albuquerque JA, Fuente C, Bernal MP. Chemical properties of anaerobic digestates affecting C and N dynamics in amended soils. Agriculture, Ecosystems and Environment. 2012;**160**:15-22. DOI: 10.1016/j.agee.2011.03.007

[69] Colmer TD, Bloom AJ. A comparison of NH_4^+ and NO_3^- net fluxes along roots of rice and maize. Plant, Cell and Environment. 1998;**21**:240-246. DOI: 10.1046/j.1365-3040.1998.00261.x

[70] Nye PH. Changes of pH across the rhizosphere induced by roots. Plant and Soil. 1981;**61**:7-26. DOI: 10.1007/BF02277359

[71] Gregory PJ, Hinsinger P. New approaches to studying chemical and physical changes in the rhizosphere: An overview. Plant and Soil. 1999;**211**:1-9. DOI: 10.1023/A:1004547401951

[72] Hinsinger P. Bioavailability of soil inorganic P in the rhizosphere as affected by root-induced chemical changes: A review. Plant and Soil. 2001;**237**:173-195. DOI: 10.1023/a:1013351617532

[73] Hinsinger P, Gobran GR, Gregory PJ, Wenzel WW. Rhizosphere geometry and heterogeneity arising from root-mediated physical and chemical processes. New Phytologist. 2005;**168**:293-303. DOI: 10.1111/j.1469-8137.2005.01512.x

[74] Devau N, Cadre EL, Hinsinger P, Jaillard B, Gérard F. Soil pH controls the environmental availability of phosphorus: Experimental and mechanistic modelling approaches. Applied Geochemistry. 2009;**24**:2163-2174. DOI: 10.1016/j.apgeochem.2009.09.020

[75] Gahoonia TS, Claassen N, Jungk A. Mobilization of phosphate in different soils by ryegrass supplied with ammonium or nitrate. Plant and Soil. 1992;**140**:241-248. DOI: 10.1007/BF00010600

[76] Kirk GJD, Santos EE, Findenegg GR. Phosphate solubilization by organic anion excretion from rice (*Oryza sativa* L.) growing in aerobic soil. Plant and Soil. 1999;**211**:11-18. DOI: 10.1023/A:1004539212083

[77] Faget M, Blossfeld S, Gillhaussen P, Schurr U, Temperton VM. Disentangling who is who during rhizosphere acidification in root interactions: Combining fluorescence with optode techniques. Frontiers in Plant Science. 2013;**4**:1-8. DOI: 10.3389/fpls.2013.00392

[78] Li L, Li SM, Sun JH, Zhou LL, Bao XG, Zhang HG, et al. Diversity enhances agricultural productivity via rhizosphere phosphorus facilitation on phosphorus-deficient soils. Proceedings of the National Academy of Sciences of the United States of America. 2007;**104**:11192-11196. DOI: 10.1073/pnas.0704591104

[79] Zhou LL, Cao J, Zhang FS, Li L. Rhizosphere acidification of faba bean, soybean and maize. Science of the Total Environment. 2009;**407**:4356-4362. DOI: 10.1016/j.scitotenv.2009.02.006

Chapter 10

Electrochemical Methods in Plant Nutritional Physiology: Ionometry and Conductometry

Alexander Kurovsky, Imran Khan and Yana Gummer

Abstract

This chapter presents an overview of the fundamental principles of the electrochemical approach to the study of plant mineral nutrition. This is particularly crucial for investigating the kinetics of macronutrient ion absorption by plant roots, alterations in leaf mineral nutrition status due to external and internal factors, and other related processes. The two principal methodological approaches to electrometry, as outlined in the section, are conductometry (the measurement of electrical conductivity in water solutions and tissue extracts) and ionometry (the determination of ions by means of ion-selective electrodes in a range of media). Each of these methodological approaches has its own set of advantages and disadvantages, as well as unique opportunities for research into the mineral nutrition of plants; however, they also present a number of inherent limitations. A solid understanding of the fundamentals of electrometry can serve as a valuable asset for researchers engaged in studies related to the mineral nutrition of plants.

Keywords: electrometry, conductometry, ionometry, ion-selective electrodes, potassium- and nitrate-selective electrodes

1. Introduction

Electrochemical research conducted in the nineteenth century demonstrated a robust correlation between the electrical properties of electrolyte solutions and their concentrations. This discovery established the foundation for the development of a method for determining solution concentration through electrical parameters. Consequently, two distinct methodologies for ascertaining ion concentration in solutions were devised: one reliant on current intensity (conductometry) and the other contingent on voltage across semipermeable membranes (potentiometry). Both methods are very important for scientific research in the area of mineral nutrition of plants. All macro- and micronutrients are absorbed by root cells mostly in ion forms. The main ions for macronutrients are nitrate and ammonium (nitrogen nutrition), potassium, calcium, magnesium ions, phosphate, and sulfate. Thus, soil solution, as well as nutrient solutions for hydroponic growing of plants, are inherently electrolyte solutions. This means that we can successfully use electrochemical methods to study these solutions and, above all, to electrochemically determine the concentration of ions in nutrient solutions (including soil solution) and tissue extracts.

This chapter is written for everyone who would like to begin mastering the methodology of fast and simple (in comparison with classical methods of analytical chemistry) electrochemical methods for the estimate of the processes and status of mineral nutrition of plants. What do you need to know to understand the basic concepts of conductometry and ionometry? To do this, it is enough to have a solid knowledge of the school course in chemistry, physics, and mathematics. For further improvement in the application of electrochemical methods of analysis in plant physiology, knowledge of a university course in general and analytical chemistry is required, as well as the study of special literature, such as, for example, see Ref. [1]. And, of course, you need to understand that no theory, no methodological manuals can make you a good specialist in the field of biological ionometry, where, as in any other area of analytical methods, there are many nuances, and many potential pitfalls. In order to understand all this, it takes years, and maybe decades of practical work, and many mistakes made, realized, and corrected. Therefore, the main goal of this chapter is to ignite a spark of interest in the researcher who wants in the future to master such an excellent tool for studying the mineral nutrition of plants as conductometry and ionometry.

2. Conductometry

This section presents a short description of the fundamental principles of the initial group of methods, namely conductometry.

Understanding these principles will help to equip students and researchers in area of the plant physiology with the primary knowledge in conducting conductometric express analyses of liquid nutrition mediums, such as different hydroponic nutritional solutions, as well as aqueous extracts from soil, greenhouse ground, and plant tissues.

What is the definition of conductometry? Conductometry is a method of measuring the concentration of ions in solutions. It is based on the dependence of the electrical conductivity of liquid media on the content of ions of various substances in them.

The term “conductometry” [2] is derived from the combination of the English word “conductivity” (denoting electrical conductivity) and the Greek word “metreo” (meaning “measure”). It has given rise to a number of related concepts, including the word “conductometer” (a device for measuring the electrical conductivity of a solution, **Figure 1**) and the term “conductometric measurements” (a method of determining the concentration of ions in solutions by conductometry).

Electrical conductivity (EC) is defined as the ability of a medium to conduct an electric current. It is important to distinguish between the electrical conductivity of solids, liquids, and gases. In the field of electrical engineering, the resistance of solid materials (both conductors and dielectrics) to the passage of electric current is a widely utilized characteristic. The unit of measurement for electrical resistance is the ohm. Electrical conductivity is inversely proportional to electrical resistance. Accordingly, the value of ohm^{-1} represents a unit of measurement for electrical conductivity, which has been designated “Siemens” or (derived from this designation) the Latin letter S.

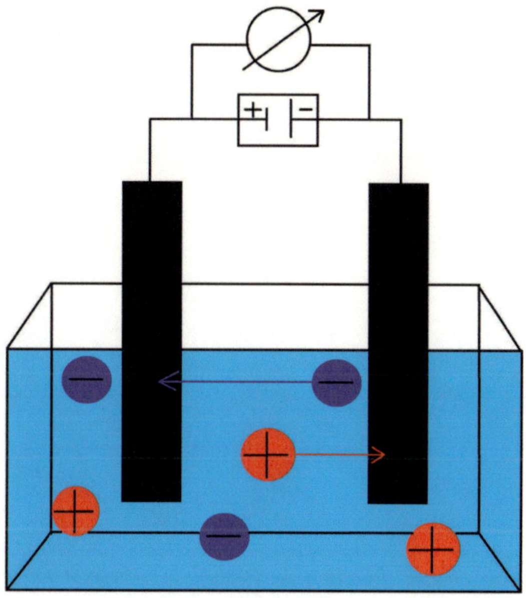

Figure 1.
The simplest scheme of the conductometric cell. The black bars denote graphite electrodes: anode (left) and cathode (right).

In the field of plant physiology, conductometry is typically employed to quantify the electrical conductivity of solutions, in which diverse physiological significant ions are present. The magnitude of the electrical conductivity of such media is relatively small; thus, the Siemens fractions are used to express it. These fractions are millisiemens and microsiemens, respectively, which correspond to 10^{-3} and 10^{-6} S or mS and mkS. As the measurements are normalized to the thickness of the liquid column in the electrode cell (**Figure 1**), the unit of measurement is expressed as $mkS \cdot cm^{-1}$.

All water-soluble substances can be classified into two principal groups: electrolytes and nonelectrolytes. The process of electrolytic dissociation refers to the decomposition of electrolyte molecules into ions. In this instance, only a proportion of the molecules undergo disintegration. The proportion of molecules that have undergone ionization is defined as the degree of dissociation, expressed as a percentage.

The electrical conductivity of solutions of different electrolytes taken in the same concentration and under the same conditions is characterized by a specific value, defined as the equivalent electrical conductivity. The equivalent electrical conductivity value is contingent upon the nature of the dissolved substance and a number of its physicochemical properties [3]. In general, all factors that affect the electrical conductivity of electrolyte solutions can be classified into three categories: specific, partially specific, and nonspecific. The initial group encompasses the genuine ion radius, the extent of dissociation, and the activity coefficient of the ion in question. The second group comprises the magnitude and sign of the charge, while the third encompasses the general parameters of the solution, including temperature and solute concentration.

Figure 2 shows the results of the calibration of a pocket conductometer Hanna Dist-3 (Netherlands) using standard water solutions of potassium chloride in the concentration range of 1–10 milliequivalents/liter (mEq/l).

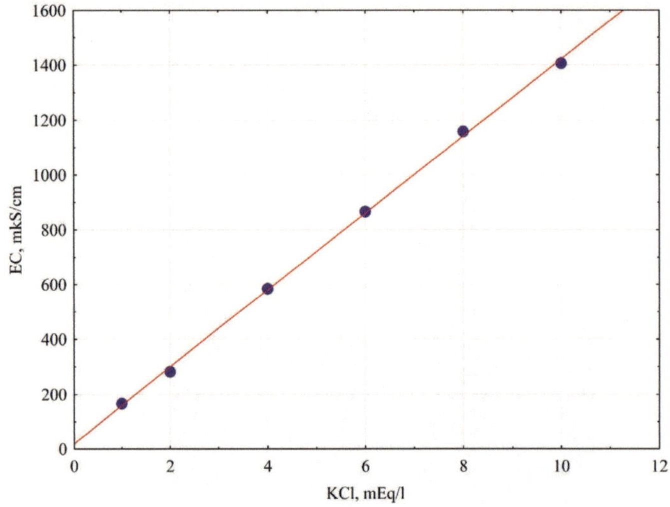

Figure 2.
The calibration graph of the dependence of electrical conductivity on the concentration of potassium chloride solutions.

The curve of dependence of electrical conductivity on concentration exhibits three distinct sections. Initially, there is a nonlinearity observed in the range of low sensitivity and noise of the device (invisible on diagrams due to small scale). This is followed by a range of linear dependence of electrical conductivity on concentration, which is then itself followed by a nonlinearity in the range of high concentrations and a subsequent decrease in ion activity [4].

As illustrated in the graph (**Figure 2**), the calibration dependence of electrical conductivity on the concentration of potassium chloride solutions is linear. It is important to note that this concentration range ensures the accuracy and reproducibility of conductometric measurements.

Devices for measuring electrical conductivity, or conductometers, can be classified into two main categories: stationary and portable. The precision of contemporary portable conductometers is adequate for utilization in biological, environmental, agrochemical, and biomedical research.

As an analytical method, conductometry has both advantages and disadvantages. The advantages of conductometry include the speed and simplicity of measurements, the stability of readings, durability, and the relative cheapness of conductometers. Besides, the sensitivity limit of the majority of conductometric measurements is 10^{-5} mol/l, and in some cases, – 10^{-6} mol/l. In the case of conductometry, the readings of detection devices are largely devoid of any discernible trend.

The main disadvantage is the non-specificity of the conductometry. This therefore precludes the possibility of determining the kind of ions in question through the use of EC measurement. A preliminary expert assessment of the composition of the samples under study can markedly enhance the informative value of conductometric analysis methods.

Thus, the combination of speed and accuracy makes conductometry an invaluable tool in many scientific disciplines. A case in point is the effective use of conductometry in plant nutrition physiology, exemplified by the conductometric monitoring of the summary concentration of ions in nutrition solutions during the hydroponic cultivation of plants. It is a highly popular and effective method for controlling growth processes.

3. Ionometry

The progression of numerous biological processes is linked to alterations in the activity of ions within the cell and in the surrounding environment. Consequently, the measurement of intracellular, extracellular, and tissue ion concentrations (e.g., H^+, K^+, etc.) can provide valuable insight into the physiological state of a living organism. At present, electrometric methods are employed extensively in numerous fields of experimental and physicochemical biology, utilizing ion-selective electrodes, which are modern electrochemical sensors [5]. These are used in conjunction with recording devices, namely potentiometers, to rapidly and specifically determine the activity of ions in solutions of diverse natures, with aqueous solutions being the most prevalent. The principal functional component of ion-selective electrodes is the ion-selective membrane (**Figure 3**), which is a thin artificial polymer film or, on occasion, a hollow glass ball. The media employed may be liquid or gaseous. The most established and renowned ion-selective electrode is the glass pH electrode, which has been in use for approximately a century. The specific properties of the membrane, regardless of its composition, are imparted by special substances that confer selective permeability to certain inorganic ions to varying degrees.

The determination method is based on the characteristic property of ion-selective membranes: formation of equilibrium potential, changing which is associated with changes in the activity of the studied ion in the solution by the Nernst eq. [5]:

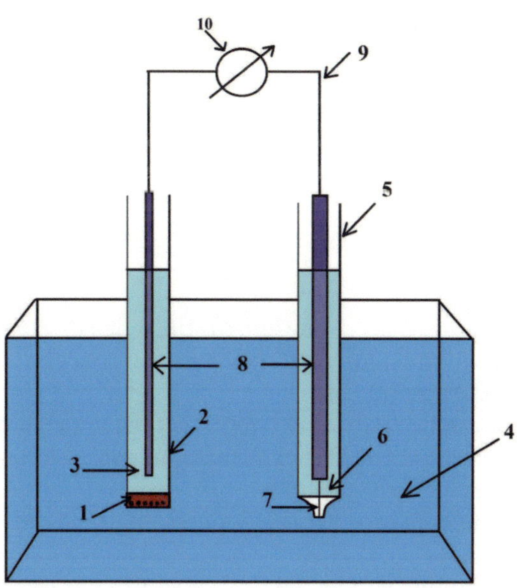

Figure 3.
The schematic diagram of an ion-selective electrode. Notes: *1—the ion-selective membrane; 2—the body of the ion-selective electrode; 3—the internal near-electrode solution; 4—the studied solution; 5—the external reference electrode; 6—the saturated solution of the potassium chloride; 7—the capillary; 8—the internal reference electrodes; 9—the metallic conductors; 10—potentiometer.*

$$E=E^0\pm\frac{RT}{nF}\ln\frac{a_1}{a_2}, \quad (1)$$

where E—is the potential of the ion-selective electrode, E_0 is the potential of the external reference electrode, R is the universal gas constant, T—absolute temperature, n—valence of the determined ion, F is the Faraday constant, a_1 and a_2—activities of the determined ion in the studied and internal near-electrode solutions, respectively, mol/l.

The "±" sign in the equation means that the potential of the ion-selective membrane changes in the opposite way for anions and cations. Many modern instruments for determining the activity of ions in aqueous solutions—ionomers, use manual or automatic mechanisms for switching polarity, which makes it possible to obtain a logically "correct" dependence when measuring both anions and cations: with an increase in the activity of the ion in a solution, the potential of the ion-selective electrode also increases.

It should be noted that there are two basic terms used in this section, characterizing the number of ions in solutions: "concentration" and "activity". They are expressed in the same units of measurement, for example in milligram-equivalents per liter (mg-eq/l, mEq/L). To clarify the difference between them, we quote from Glinka's textbook on general chemistry: "The activity of an ion is understood as that effective, conditional concentration of it, according to which it acts in chemical reactions" [6].

The relationship between concentration and activity is expressed by the equation:

$$a_x=\gamma\times C_x, \quad (2)$$

where a_x is the activity of the ion in the solution, C_x is the concentration of the ion in the solution, and γ is the activity coefficient of the ion.

In highly dilute solutions, the activity coefficient tends to be 1, and the activity itself is practically equal to the concentration. In more concentrated solutions of electrolytes, ion activity coefficients are noticeably less than 1. A decrease in the activity coefficient can be due to various reasons, but the main parameter affecting the activity coefficient is the so-called ionic strength of a solution, which is a measure of the intensity of the electric field created by ions in a solution and is calculated by the equation:

$$I=\frac{1}{2}\sum_{i=1}^{n}C_i\times Z_i^2, \quad (3)$$

where I is the ionic strength of a solution, i is the index of the specific ion, n is the number of specific ions in the solution, C_i is the partial concentration of the specific ion, Z_i—the charge of the specific ion.

Based on Eq. (3), it can be easily understood that the ionic power of mono-solutions of salts, which consist of monovalent anion and cation, is equal to the molar concentration of these solutions.

Table 1 illustrates the impact of varying ionic strength on the activity coefficients of K^+ and Cl^-.

In the physiological studies conducted the ionometric measurements most often in the concentration range of the determined ions 0.001–0.01 mol/l. At the same time, many other ions make a significant contribution to the values of ion strength of solutions, because real biological solutions (for example, aqueous extracts of plant tissues) have, as a rule, a multicomponent electrolyte composition. How, by measuring

I	0.0005	0.001	0.005	0.01	0.05	0.1
γ	0.975	0.964	0.925	0.899	0.805	0.755

Table 1.
The values of the activity coefficients of K^+ and Cl^- depending on the ionic strength of solutions. Legend is the same as in Eq. (3).

the activity of ions with the help of potentiometry, can we determine, ultimately, their concentration?

The initial step is to identify the most suitable dilutions of the solutions under investigation. The concept of "optimal breeding" implies that the concentration of the determined ion should be, on the one hand, significantly higher than the sensitivity threshold of modern ionomers (for most electrodes, the sensitivity threshold is 10^{-5}–10^{-6} mol/l), and on the other hand, not result in ranges where the coefficient of activity of the determined ion is significantly lower than one. As evidenced by practice, such optimal ranges of the studied solutions correspond to the range of concentrations of most ion-macronutrients, 10^{-3}–10^{-2} mol/l.

Secondly, the standard solutions (used to calibrate ionomers) are necessary, as accurately as possible corresponding with the studied samples according to the parameter of ionic strength. To do this, usually in a mono-solution of the salt containing the determined ion, "load" electrolytes are added, which provide the required ionic strength. For example, when measuring the concentration of potassium ions in multicomponent biological fluids, a certain number of other salts (sodium chloride, calcium chloride, sodium acetate, etc.) are added to standard potassium chloride solutions.

Thirdly, the concentrations of the determined ion in standard solutions should be selected in such a way that the studied (unknown) samples should fall into the range between the extreme calibration points. For example, if when calibrating the potassium-selective electrode the standard solutions with concentrations of 1 and 10 mEq/l are used, then it is necessary to achieve such an unknown sample so that the concentration of potassium ions in it is in the range of 1–10 mEq/l.

If all the above conditions are met, measuring the activity of the determined ion in a solution with a high degree of accuracy will give us the concentration values of this ion.

It is beneficial to consider the terminology used to describe instruments in ionometry in greater detail. In certain instances, one may encounter the following designations: "millivoltmeter," "potentiometer," "pH meter," and "ionomer." It is our contention that a distinction must be drawn between these terms and that a hierarchy must be established.

A millivoltmeter is a mandatory measuring instrument, an internal device for recording the potential difference (membrane potential) arising on the electrodes in the measuring cell. Maybe in the analog and in the digital construction.

A potentiometer is a device for measuring potential (more precisely, electromotive force, EMF) in electrochemical cells of various designs, which, in addition to a millivoltmeter, contains other structural elements, such as a signal amplifier, calibration, and sensitivity adjustment resistors, in some cases—recorders, etc.

An Ionometer represents a specific type of potentiometer. Ionometers are used to determine the activity (concentration) of ions in water solutions by means of ion-selective electrodes. A pH meter represents a specific category of ion meter. This instrument is employed in the determination of hydrogen ion activity through the utilization of a glass pH electrode.

The pH index introduced more than a century ago, subsequently gave rise to a more general designation widely used in electrochemistry—pX (Eq. [4]).

$$pX=-lga_x, \quad (4)$$

where a_x is the activity of any ion in solution (not just H^+), mol/l.

In specific cases, we can say about such indexes, as pK^+, pNO_3^-, pCa^{2+}, etc.

The general approach when calibrating instruments used in ionometric measurements consists of taking readings from the potentiometer (the values of the potential in millivolts, mV) while sequentially immersing the electrode system in a series of standard solutions with given, previously known values of pX.

Figure 4 illustrates the calibration outcomes of the Ionometer IPL-103 (Russia) utilizing standard water solutions of potassium chloride in a concentration range of 1–10 mEq/l (pX range 2–3).

The following section presents a general algorithm for measuring the activity of an ion in a studied sample, along with the subsequent calculations.

1. The electrode system is immersed in the test solution, which contains an unknown concentration of the ion that is to be determined.

2. The potentiometer readings are duly recorded.

3. The potential values of the ion-selective electrode are converted into pX values.

4. The values of pX are then recalculated into activity values in accordance with the specified Eq. (5).

$$a_x=10^{-pX}, \quad (5)$$

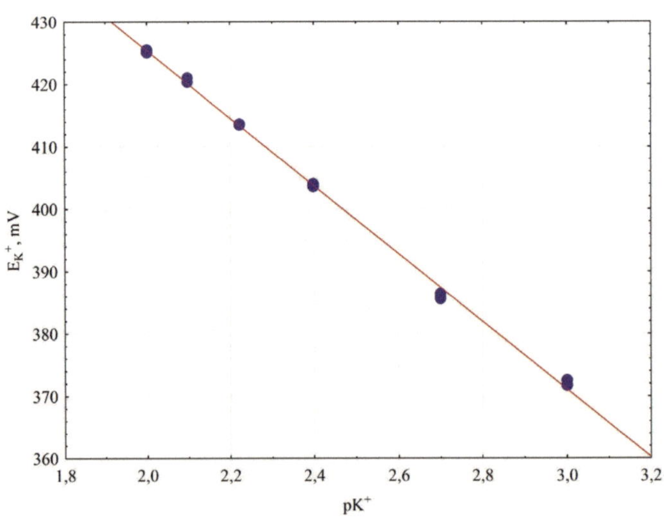

Figure 4.
The calibration dependence of the potential of the potassium-selective electrode (E_K^+) on the pK^+ – value in the standard solution.

The legend is identical to that presented in the Eq. (4).

Let us look at the advantages and disadvantages associated with the use of ionometry as an analytical method in the field of plant physiology.

The method's indisputable advantages include the selectivity of the electrodes. This allows us to ascertain the concentration (activity) of a particular ion present in the solution under analysis. Ion-selective electrodes are a common tool for measuring the activity of specific ions, including potassium (K^+), nitrate (NO_3^-), ammonium (NH_4^+), calcium (Ca_2^+), and chloride (Cl^-). Except chloride anion, all of these elements are essential macronutrients for the mineral nutrition of plants.

Ionometric measurements are a rapid and cost-effective method of measurement that does not necessitate the use of expensive equipment.

The use of ion-selective electrodes enables the observation of specific ion absorption processes by plant roots in real-time within a hydroponic system.

The concentration of specific macronutrient ions in plant tissues allows for the preparation of aqueous extracts for ionometric measurements with the most convenient dilution. This is particularly the case with potassium ions, which have a concentration of approximately 10^{-2} mol/l in plant cells [7].

The disadvantages of ion-selective electrodes are a consequence of the aforementioned advantages.

It is evident that ion-selective membranes lack the requisite degree of selectivity for certain ions. Based on our experience, ion-selective electrodes are the most reliable for measuring the activity of K^+ and NO_3^-. However, even when utilizing these electrodes, it is imperative to consider the potential for interference from other ions present in the solution, as they can impact the electrode potential and introduce inaccuracies in the measurement outcome.

It is not feasible to continuously monitor the concentration of specific ions in hydroponic solutions over an extended period of time. This is due to the inherent limitations of the external reference electrode, which cannot maintain a constant potential for an extended period, such as several hours. Periodic recalibration of the electrode system is necessary, which may be perceived as an inconvenience.

Ionometry is distinguished by a relatively high degree of measurement uncertainty. The relative uncertainty of measurements using ion-selective electrodes in multicomponent media (e.g., in extracts from plant tissues) is approximately 10–15% for a concentration of the determined ion of around 10^{-2} mol/l, about 20% for a concentration of 10^{-3} mol/l, 25–30% or more for a concentration of 10^{-4} mol/l and below.

While there are a number of potential disadvantages and limitations to consider, ion-selective electrodes have the potential to be a valuable and effective research tool in the field of studying the physiology of plant mineral nutrition and agrochemical research.

We employ ionometry techniques with the aim of studying the mineral nutrition status of plants and the processes of extracting macronutrient ions from experimental nano-fertilizers [8–10].

As an example of research conducted using the methods described in this chapter, we can cite our experiments devoted to the study of woody leaf litter processed in vermiculture as a source of macronutrients for the mineral nutrition of plants. The general ideology and methodology (including the methods of water extraction of ions from substrates, etc.) of such studies can be understood from the papers [11, 12]. However, for practical confirmation of everything stated in this chapter, we will cite hitherto unpublished data from recent experiments.

Potassium form	Stages of the vermicomposting		
	1	2	3
Total	1.02 ± 0.05	1.01 ± 0.05	1.07 ± 0.1
Water-soluble K^+	0.23 ± 0.02	0.42 ± 0.02	0.48 ± 0.03

Table 2.
The concentration of potassium (percent of dry mass) in poplar leaf litter processing in vermiculture. Notes: *1–the start of the vermicomposting (unprocessed leaf litter); 2—finish of the vermicomposting (5 months later); 3—storage and "maturation" of the raw vermicompost in sealed polyethylene bags (17 months after finish); the data in the means (of measurement series) with the 95% confidence interval are presented in the table.*

We studied the content of one of the most important macronutrients—potassium, in poplar leaf litter before and after its vermicomposting. One of the goals of this work was to assess the possibility of using poplar leaf litter as a component of biofertilizers, the introduction of which would reduce the damage from the removal of macronutrients from urban and agroecosystems. We made successive measurements of the potassium content in the leaf litter using two methods: the classical method of the total potassium determination (the flame photometry after combustion of a sample in a mixture of the concentrated sulfuric acid and hydrogen peroxide) and ionometry using potassium-selective electrodes (the water-soluble K^+ determination). In total, there were three stages of leaf litter sampling and subsequent measurements of the potassium concentration: at the start of vermicomposting, upon its completion (after 5 months), and after almost a year and a half of "maturation" of raw vermicompost in sealed polyethylene bags.

Table 2 shows the results of measurements of the potassium concentration in the processing of poplar leaf litter in vermiculture.

The table clearly shows that the total potassium content in the processing substrates does not change over time. This indicator serves only to assess the total potential reserves of potassium in the biofertilizer. At the same time, the data from ionometric measurements are valuable in that they demonstrate the dynamics of the increase in the content of water-soluble potassium in the substrate as it is processed. It can be said that the two methods presented in the example, on the one hand, solve different analytical problems, and on the other hand, complement each other well.

4. Conclusion

It is widely acknowledged that conductometry is an excellent method for analyzing the dynamics of plant absorption of macronutrient ions from mono-solutions of salt (for example KNO_3, NH_4NO_3, $CaCl_2$, etc.). It is worth noting that the sensitivity limit of modern conductometers is 1–2 mkS/cm. This means that you can be reassured that you can analyze very dilute solutions down to distilled water. It is worth noting that conductometers are largely free of any drift in their readings over time.

Ionometry represents the most effective method of analysis for multicomponent samples and therefore represents a valuable tool in this context. As an example, one might consider the quantification of the content of potassium, calcium, and nitrates in the leaf of the plant under study. In order to achieve this, it would be advisable to conduct ionometric measurements in aqueous extracts from plant leaves. It may be beneficial to consider increasing the number of statistical replicates for each sample tested in order to reduce measurement uncertainty in the area of low ion concentrations.

It is beyond doubt that conductometry and ionometry are analytical methods that, when employed correctly, can be of significant value in the field of plant mineral nutrition physiology.

Author details

Alexander Kurovsky*, Imran Khan and Yana Gummer
National Research Tomsk State University, Tomsk, Russian Federation

*Address all correspondence to: a.kurovskii@yandex.ru

References

[1] Bagotsky VS. Fundamentals of Electrochemistry. New Jersey: Wiley; 2005. 725p. DOI: 10.1002/047174199X

[2] Malhotra P. Conductometric methods. In: Analytical Chemistry. Cham: Springer; 2023. DOI: 10.1007/978-3-031-26757-4_11

[3] Tikhonov I. The Influence of Ion Composition of Water on its Electrical Conductivity [Internet]. 2024. Available from: https://tiwater.info/en/the-influence-of-ion-composition-of-water-on-its-electrical-conductivity/ [Accessed: August 12, 2024]

[4] Klein KA, Santamarina JC. Electrical conductivity in soils: Underlying phenomena. JEEG. 2003;**8**:4. DOI: 10.4133/JEEG8.4.263

[5] Zhai J, Luo B, Li A, Dong H, Jin X, Wang X. Unlocking all-solid ion selective electrodes: Prospects in crop detection. Sensors. 2022;**22**:5541. DOI: 10.3390/s22155541

[6] Glinka NL. General Chemistry. Vol. 1. Moscow: Mir Publishers; 1986. 372p

[7] Clarkson DT. Ion Transport and Cell Structure in Plants. New Jersey: Wiley; 1974. 350p

[8] Kurovsky AV, Burenina AA, Mikhaylova SI, Petrochenko KA, Astafurova TP. Amaranth nutritional properties assessment based on potassium and nitrate concentration in tissues. BBRA. 2015;**12**(3):2161-2166. DOI: 10.13005/bbra/1887

[9] Rudmin M, Makarov B, López-Quirós A, Maximov P, Lokteva V, Ibraeva K, et al. Preparation, features, and efficiency of nanocomposite fertilizers based on Glauconite and ammonium dihydrogen phosphate. Materials. 2023;**16**:6080. DOI: 10.3390/ma16186080

[10] Rudmin M, Maximov P, Dasi E, Kurovsky A, Gummer Y, Ibraeva K, et al. Intercalation of carbamide to globular glauconite by chemical processing for the creation of slow-release nanocomposites. Applied Clay Science. 2023;**243**:107075. DOI: 10.1016/j.clay.2023.107075

[11] Petrochenko K, Kurovsky A, Babenko A, Yakimov Y. Leaf litter-based vermicompost as promising calcium fertilizer. Tomsk State University Journal of Biology. 2015;**2**(30):20-34. DOI: 10.17223/19988591/30/2

[12] Kurovsky A, Petrochenko K, Godymchuk A, Babenko A, Yakimov Y. Physicochemical aspects of recycling tree leaf litter in the south of Western Siberia by the Eisenia fetida (Savigny) vermiculture. In: IOP Conference Series: Earth and Environmental Science, Volume 226, International Jubilee Scientific and Practical Conference "Innovative Directions of Development of the Forestry Complex". Voronezh, Russian Federation; 2018. DOI: 10.1088/1755-1315/226/1/012009